RAND

Trends in the Global Balance of Airpower

Supporting Data

Christopher J. Bowie, Kirninder Braich,
Lory Arghavan, Marcy Agmon,
Mary Morris

Prepared for the
United States Air Force

Project AIR FORCE

Preface

This document contains the tabular data referred to in the report *Trends in the Global Balance of Airpower*, MR-478/1-AF. It provides a country-by-country accounting of military aircraft, surface-to-air missile (SAM), and helicopter inventories for the year 1991. It also provides the economic data on individual nations employed in the analysis for the years 1979–1989.

Project AIR FORCE

Project AIR FORCE, a division of RAND, is the Air Force federally funded research and development center (FFRDC) for studies and analyses. It provides the Air Force with independent analyses of policy alternatives affecting the development, employment, combat readiness, and support of current and future aerospace forces. Research is being performed in three programs: Strategy, Doctrine, and Force Structure; Force Modernization and Employment; and Resource Management and System Acquisition.

Project AIR FORCE is operated under Contract F49620-91-C-0003 between the Air Force and RAND.

Contents

1. Helicopter Data

The following table provides data on individual national inventories of military helicopters. We have specified general roles of the various assets as follows:

- Utility helicopters are typically not equipped with armament (with perhaps the occasional exception of a few light defensive guns). They tend to be the larger helicopters and are employed in such operations as airlift, medical evacuation, fire control, support, survey, search and rescue, utility, light observation, minesweeping, surveillance, and special operations.

- Armed helicopters are employed in land and sea combat. Armed helicopters designed for land operations can be armed with guided missiles, heavy cannon, and rockets; those configured for maritime operations can be armed with a mix of guided missiles and heavy cannon (for attacks against surface vessels) and/or torpedoes and depth charges (for attacks against submarines).

- Helicopters labeled as "unknown" were included in the utility category in the report's figures.

Country	Number	Name	Role
Afghanistan			
	25	Mil Mi-8 'Hip E'	Armed
	30	Mil Mi-25 'Hind' (export Mi-24D)	Armed
	35	Mil Mi-17 'Hip H'	Armed
	12	Mil Mi-4 'Hound'	Utility
Albania			
	28	Harbin Z-5 (Mi-4)	Armed
	20	Pinkiang H-4 (Mil Mi-4) 'Hound'	Utility
Algeria			
	40	Mil Mi-25 'Hind' (export Mi-24D)	Armed
	6	ICA-Brasov/Aerospatiale IAR-316B Alouette III	Armed
	5	Mil Mi-6 'Hook'	Armed
	29	Mil Mi-8/-17 'Hip D/H'	Armed
	6	Bredanardi/Schweizer-Hughes 300C	Utility
	38	Mil Mi-24 'Hind'	Utility
	25	Mil Mi-4 'Hound'	Utility
Angola			
	6	Aerospatiale SA-365M Dauphin II (SA-360)	Armed
	6	Eurocopter (Aerospatiale) SA-342M Gazelle	Armed
	37	Mil Mi-25 'Hind' (export Mi-24D)	Armed
	30	ICA-Brasov/Aerospatiale IAR-316B Alouette III	Armed
	17	Mil Mi-17 'Hip H'	Armed
	50	Mil Mi-8 'Hip C'	Armed
	32	Aerospatiale SA-316A/B Alouette III (SA-319/SE-3160)	Utility
	10	Aerospatiale SA-365C Dauphin II (SA-360)	Utility
	12	Aerospatiale/Brasov SA-316M	Utility
	1	Aerospatiale SA-315 Lamma (Alouette II)	Unknown
	15	Aerospatiale SA-365AA/UA Panther	Unknown
Argentina			
	7	Aerospatiale Alouette III SA-319B	Armed
	3	Aerospatiale AS-332B Super Puma	Armed
	3	Aerospatiale/Westland SA-330 Puma	Armed
	3	Aerospatiale/Westland SA-330 Puma	Armed
	6	MD 500MD Defender II	Armed
	11	MD 500MD Defender II	Armed
	4	Sikorsky/Agusta ASH-3H Sea King	Utility
	10	Agusta A-109 Hirundo	Utility
	21	Bell (205) UH-1H Iroquois	Utility
	5	Bell 205 (CH-118) Iroquois	Utility
	12	Bell 205A-1/UH-1H Iroquois	Utility
	8	Bell 212 (CH-135) Twin-Pac/Twin Huey	Utility
	5	Boeing CH-47C Chinook (CH-147)	Utility
	24	Eurocopter (Aerospatiale) A-532UC Super Puma	Utility
	4	HAL/Aerospatiale SA-315B Cheetah	Utility
	8	Hiller UH-12ET (OH-23)	Utility
	2	IAR/Aerospatiale IAR-315 (SA-315B)	Utility
	2	Sikorsky UH-34T (S-58T)	Unknown
	4	Aerospatiale SA-315 Lamma (Alouette II)	Unknown
	5	Aerospatiale SA-315 Lamma (Alouette II)	Unknown

Country	Number	Name	Role
Argentina	4	Fairchild-Hiller FH-1100 (OH-5)	Unknown
	4	MD Hughes 500D Defender	Unknown
	2	S-61R/NR	Unknown
	4	Sikorsky/Agusta AS-61D (SH-3)	Unknown
	2	Westland Lynx HAS. Mk23	
Australia			
	16	Sikorsky (SH-60) S-70B-2 Seahawk	Armed
	24	Aerospatiale AS-350B Ecureuil	Armed
	12	Westland Sea King Mk 50A	Armed
	2	BAe-748	Utility
	25	Bell (205) UH-1H Iroquois	Utility
	12	Boeing CH-47C Chinook (CH-147)	Utility
	46	CAC/Bell 206B-1 Kalkadoon (Jetranger)	Utility
	24	Eurocopter (Aerospatiale) AS-350 Ecureuil	Utility
	34	Sikorsky S-70 (UH-60)	Utility
	39	Sikorsky S-70A-9 Black Hawk	Utility
Austria			
	25	Aerospatiale SA-316B Alouette III (SE-3160)	Utility
	17	Agusta/Bell 204B Iroquois	Utility
	23	Bell 212 (CH-135) Twin-Pac/Twin Huey	Utility
	12	Bell CH-136 (OH-58) Kiowa	Utility
Bahrain			
	2	Bell Military 412	Armed
	4	MBB Bo-105C (NBo-105)	Armed
	12	Bell (205) UH-1H Iroquois	Utility
	2	Sikorsky (UH-60L) S-70B Black Hawk (SH-60)	Utility
Bangladesh			
	6	Mil Mi-8 'Hip C'	Armed
	10	Bell 212 (CH-135) Twin-Pac/Twin Huey	Utility
	3	Bell UH-1N (Bell 212)	Utility
Belgium			
	28	Agusta A.109BA	Armed
	18	Agusta A.109CM Naval	Armed
	5	Westland Sea King Mk 48	Armed
	59	Aerospatiale Alouette II/Astazou (SE-3130) SA-318C	Utility
	4	Aerospatiale SA-316B Alouette III (SE-3160)	Utility
Benin			
	2	Aerospatiale AS-350B Ecureuil	Armed
	1	Kamov Ka-26 'Hoodlum'	Armed
	2	Aerospatiale Alouette II/Astazou (SE-3130) SA-318C	Utility
	3	Aerospatiale SA-350B/355M Ecureuil 1/2	Unknown
	1	SA-355M	Unknown
Bolivia			
	27	Bell (205) UH-1H Iroquois	Utility
	2	Bell 212 (CH-135) Twin-Pac/Twin Huey	Utility
	6	Helibras/Aerospatiale HB-315B Gaviao (SA-315)	Utility
	2	IAR/Aerospatiale IAR-315 (SA-315B)	Utility
	1	UH-1 Iroquois (Bell 204/205)	Utility
Bophuthatswana			
	1	HAL/Aerospatiale SA-315B Cheetah	Utility

Country	Number	Name	Role
Botswana			
	2	Aerospatiale AS-350L1 Ecureuil	Armed
	5	Bell Military 412	Armed
Brazil			
	7	Eurocopter (Aerospatiale) AS-532SC Cougar	Armed
	6	Westland Wasp HAS-1	Armed
	6	Aerospatiale AS-332F Super Puma Naval	Armed
	6	Aerospatiale/Westland SA-330 Puma	Armed
	9	Eurocopter (Aerospatiale) AS-332 Super Puma	Armed
	11	Helibras/Aerospatiale UH-12B (Naval HB 355F2) Esquilo	Armed
	15	Sikorsky/Agusta ASH-3H Sea King	Armed
	36	Aerospatiale SA-365C Dauphin II (SA-360)	Utility
	53	Bell (205) UH-1H Iroquois	Utility
	24	Eurocopter (Aerospatiale) AS-355 Ecureuil II Twinstar	Utility
	3	Helibras/Aerospatiale HB-315B Gaviao (SA-315)	Utility
	16	Helibras/Aerospatiale HB-350[CH-50] Esquilo (AS-350)	Utility
	2	Helibras/Aerospatiale VH-55 (CH-55) Esquilo	Utility
	4	Hughes OH-6A Cayuse	Utility
	15	TH-57 Sea Ranger (Bell 206)	Utility
	30	Aerospatiale SA-350B Ecureuil	Unknown
	25	Bell OH-13E/G/H Sioux	Unknown
	6	Bell SH-1H	Unknown
	8	Helibras/Aerospatiale HB-350B/B1 Esquilo (AS-350B)	Unknown
	13	Helibras/Eurocopter (Aerospatiale) AS-555 Fennec (CH-55)	Unknown
	9	Westland Lynx HAS. Mk21	Unknown
Brunei			
	7	MBB Bo-105CBS (NBo-105)	Armed
	12	Bell 212 (CH-135) Twin-Pac/Twin Huey	Utility
	3	Bell 214ST SuperTransport	Utility
	2	Sikorsky S-70A-14 Black Hawk	Utility
Bulgaria			
	21	PZL Swidnik (Mil) Mi-2URP Hoplite	Armed
	12	Mil Mi-14PL 'Haze A'	Armed
	5	Kamov Ka-25 'Hormone'	Armed
	25	Mil Mi-17 'Hip'	Armed
	45	Mil Mi-24 'Hind F'	Armed
	26	Mil Mi-4 'Hound'	Utility
Burkina Faso			
	2	Aerospatiale SA-316B Alouette III (SE-3160)	Utility
	2	Aerospatiale SA-365N Dauphin II (SA-360)	Utility
Burma			
	20	PZL Swidnik W-3 Sokol (Falcon)	Armed
	9	Aerospatiale SA-316 Alouette III (SA-319)	Utility
	8	Aerospatiale SA-316B Alouette III (SE-3160)	Utility
	10	Bell (205) UH-1H Iroquois	Utility
	12	Bell 205 (CH-118) Iroquois	Utility
	2	Kawasaki/Vertol KV-107-II (CH-46)	Utility
	10	Mil Mi-2 'Hoplite'	Unknown
Burundi			
	4	Aerospatiale SA-342L Gazelle	Armed
	3	Aerospatiale SA-316B Alouette III (SE-3160)	Utility

Country	Number	Name	Role
Cambodia			
	10	Mil Mi-17 'Hip H'	Armed
	4	Mil Mi-8 'Hip C'	Armed
	3	Mil Mi-24 'Hind'	Utility
Cameroon			
	4	Aerospatiale SA-342L Gazelle	Armed
	2	Aerospatiale/Westland SA-330 Puma	Armed
	1	Eurocopter (Aerospatiale) AS-332 Super Puma	Armed
	3	Aerospatiale Alouette II/Astazou (SE-3130) SA-318C	Utility
	4	Aerospatiale Alouette II/Astazou (SE-3130) SA-318C	Utility
	5	Aerospatiale SA-316 Alouette III (SA-319)	Utility
Canada			
	50	Westland/Agusta EH-101 Merlin	Armed
	9	Bell 205 (CH-118) Iroquois	Utility
	50	Bell 212 (CH-135) Twin-Pac/Twin Huey	Utility
	130	Bell CH-136 (OH-58) Kiowa	Utility
	14	Bell CH-139 (Bell 206) Kiowa III	Utility
	9	Boeing CH-47C Chinook (CH-147)	Utility
	14	Boeing-Vertol CH-46 Sea Knight (CH-113A Voyageur)	Utility
	7	Boeing-Vertol CH-47-234 Chinook (CH-147)	Utility
	35	Sikorsky CH-124A/U (SH-3, CHSS-2) Sea King	Utility
Central African Republic			
	1	Aerospatiale Alouette II/Astazou (SE-3130) SA-318C	Utility
	1	Aerospatiale SA-313B Alouette II	Utility
	1	Eurocopter (Aerospatiale) AS-350 Ecureuil	Utility
Chad			
	4	Aerospatiale/Westland SA-330B Puma	Armed
	1	Aerospatiale SA-341 Gazelle	Utility
Chile			
	4	Eurocopter (Aerospatiale) AS-532SC Cougar	Armed
	4	Aerospatiale AS-365F Dauphin II	Armed
	5	Eurocopter (Aerospatiale) AS-332 Super Puma	Armed
	4	Eurocopter (Aerospatiale) AS-565SA Panther	Armed
	12	MBB Bo-105 (NBo-105)	Armed
	6	MBB Bo-105CBS (NBo-105)	Armed
	4	Aerospatiale AS-332L Super Puma	Utility
	5	Aerospatiale SA-313B Alouette II	Utility
	8	Aerospatiale SA-316 Alouette III (SA-319)	Utility
	9	Aerospatiale/Westland SA-330F Puma	Utility
	10	Bell (205) UH-1H Iroquois	Utility
	15	Enstrom 280FX	Utility
	4	Eurocopter AS-565MA Panther	Utility
	5	IAR/Aerospatiale IAR-315 (SA-315B)	Utility
	1	Sikorsky S-76 Shadow	Utility
	10	Aerospatiale SA-315 Lama (Alouette II)	Unknown
	2	Agusta-Bell 206	Unknown
	3	Agusta-Bell 206	Unknown
	10	MD Hughes 530F Defender	Unknown

Country	Number	Name	Role
China			
	340	Harbin Z-5 (Mi-4)	Armed
	30	Harbin Z-9 (Aerospatiale SA-365 Dauphin)	Armed
	60	Harbin Z-9 (Aerospatiale SA-365 Dauphin)	Armed
	15	Aerospatiale SA-321 Super Frelon	Armed
	8	Aerospatiale SA-342L Gazelle	Armed
	15	Changhe CAF Z-8 (SA-321)	Armed
	6	Eurocopter (Aerospatiale) AS-332 Super Puma	Armed
	100	Harbin Z-6 (Z-5)	Armed
	24	Mil Mi-17 'Hip'	Armed
	30	Mil Mi-8 'Hip C'	Armed
	4	Bell 214	Utility
	24	Sikorsky S-70C-2	Utility
	23	Sikorsky S-70C-2 Hei Ying (Black Hawk)	Utility
	2	Sikorsky S-76C	Utility
Ciskei			
	1	MBB Bo-105 (NBo-105)	Armed
Columbia			
	2	Bell Military 412	Armed
	4	MBB Bo-105 (NBo-105)	Armed
	20	Bell 205A-1/UH-1H Iroquois	Utility
	4	Bell 212 (CH-135) Twin-Pac/Twin Huey	Utility
	8	Bell 47G (OH-13)	Utility
	8	Hughes OH-6A Cayuse	Utility
	2	Hughes TH-55A Osage	Utility
	24	MD 500MG Night Fox	Utility
	5	MD Hughes 300C	Utility
	5	Sikorsky (UH-60A) S-70B Black Hawk (SH-60)	Utility
	9	Sikorsky (UH-60A) S-70B Black Hawk (SH-60)	Utility
	6	TH-55 Hughes 269	Unknown
Congo			
	2	Aerospatiale Alouette II/Astazou (SE-3130) SA-318C	Utility
	2	Aerospatiale SA-316 Alouette III (SA-319)	Utility
	1	Aerospatiale SA-360 Dauphin	Utility
	1	Aerospatiale SA-365C Dauphin II (SA-360)	Utility
Costa Rica			
	1	Fairchild-Hiller FH-1100 (OH-5)	Unknown
	2	MD Hughes 500D Defender	Unknown
Cote D'Ivoire			
	2	Aerospatiale/Westland SA-330 Puma	Armed
	1	Aerospatiale Alouette II/Astazou (SE-3130) SA-318C	Utility
	1	Aerospatiale SA-316 Alouette III (SA-319)	Utility
	4	Aerospatiale SA-365C Dauphin II (SA-360)	Utility
Cuba			
	30	Mil Mi-25 'Hind' (export Mi-24D)	Armed
	14	Mil MI-14PW 'Haze'	Armed
	40	Mil Mi-17 'Hip H'	Armed
	10	Mil Mi-8 'Hip C'	Armed
	60	Mil Mi-8/-17 'Hip'	Armed
	30	Mil Mi-1 'Hare'	Utility
	55	Mil Mi-4 'Hound'	Utility
	2	WSK/Mil Mi-2 'Hoplite'	Unknown

Country	Number	Name	Role
Cyprus			
	4	Aerospatiale SA-342L Gazelle	Armed
Czechoslovakia			
	56	Mil Mi-24W 'Hind E'	Armed
	52	PZL Swidnik (Mil) Mi-2RM/URP/US Hoplite	Armed
	50	Mil Mi-17 'Hip H'	Armed
	52	Mil Mi-8 'Hip C'	Armed
	60	Mil Mi-1 'Hare'	Utility
	4	Mil Mi-8 'Hip K'	Utility
	1	Mil Mi-9 'Hip G'	Utility
Denmark			
	12	Eurocopter (Aerospatiale) AS-550C2 Fennec/TOW	Armed
	2	Westland Lynx Mk 90	Armed
	8	Sikorsky S-61/SH-3 Sea King	Armed
	8	Westland Lynx Mk 80	Armed
	14	MD Hughes 500M/OH-6	Unknown
Djibouti			
	3	Aerospatiale AS-355F1 Fennec	Armed
	2	Aerospatiale Alouette II/Astazou (SE-3130) SA-318C	Utility
Dominican Republic			
	2	Aerospatiale Alouette II/Astazou (SE-3130) SA-318C	Utility
	1	Aerospatiale SA-360 Dauphin	Utility
	1	Aerospatiale SA-365C Dauphin II (SA-360)	Utility
	8	Bell 205A Iroquois	Utility
	1	Hughes OH-6A Cayuse	Utility
	1	Hughes 500D	Unknown
Ecuador			
	13	Aerospatiale AS-332B Super Puma	Armed
	4	Aerospatiale AS-350B Ecureuil	Armed
	3	Aerospatiale/Westland SA-330 Puma	Armed
	1	Aerospatiale/Westland SA-330L Puma	Armed
	8	Eurocopter (Aerospatiale) AS-332 Super Puma	Armed
	30	Aerospatiale Gazelle (SA-342K/L)	Utility
	6	Aerospatiale SA-316B Alouette III (SE-3160)	Utility
	2	Bell (204) UH-1B Iroquois	Utility
	24	Bell (205) UH-1H Iroquois	Utility
	1	Bell 212 (CH-135) Twin-Pac/Twin Huey	Utility
	1	Bell 214B Big Lifter	Utility
	3	IAR/Aerospatiale IAR-315 (SA-315B)	Utility
	3	Aerospatiale SA-315B Lama	Unknown

Country	Number	Name	Role
Egypt			
	10	HAL/Aerospatiale SA-316B Alouette III/Chetak	Armed
	6	Mil Mi-6 'Hook'	Armed
	50	Mil Mi-8 'Hip C'	Armed
	6	Westland Sea King Mk 47	Armed
	15	Agusta/Boeing CH-47C	Utility
	18	Hiller UH-12E (OH-23)	Utility
	12	Mil Mi-4 'Hound'	Utility
	4	Sikorsky (UH-60L) S-70B Black Hawk (SH-60)	Utility
	2	Sikorsky S-70A-21 Black Hawk	Utility
	25	Sikorsky/Agusta AS-61 (SH-3)	Utility
	28	Westland Commando Sea King HC.Mk4	Utility
	74	Aerospatiale SA-342K/L Gazelle	Unknown
	12	Aerospatiale/SOKO Gazela (SA-342)	Unknown
El Salvador			
	5	Aerospatiale SA-316B Alouette III (SE-3160)	Utility
	35	Bell (205) UH-1H Iroquois	Utility
	15	Bell UH-1M Iroquois	Utility
	6	Schweizer-Hughes 300C	Utility
	3	Aerospatiale SA-315B Lama	Unknown
Ethiopia			
	10	HAL/Aerospatiale SA-316B Alouette III/Chetak	Armed
	10	ICA-Brasov/Aerospatiale IAR-316B Alouette III	Armed
	2	Mil Mi-14 'Haze'	Armed
	30	Mil Mi-8 'Hip C'	Armed
	1	IAR (Eurocopter France) IAR-330L Puma	Utility
	23	Mil Mi-24 'Hind'	Utility
	1	IAR/Aerospatiale IAR-330 (SA-330)	Unknown
	3	UH-1 Iroquois (Bell 204/205)	Unknown
Fiji			
	1	Aerospatiale AS-355F2 Fennec	Armed
	1	Aerospatiale SA-365C Dauphin II (SA-360)	Utility
Finland			
	3	Eurocopter (Aerospatiale) AS-332 Super Puma	Armed
	7	Mil Mi-8 'Hip P/T'	Unknown

Country	Number	Name	Role
Former Soviet Union			
	21	Mil Mi-24P 'Hind F'	Armed
	1740	Mil Mi-24W 'Hind E'	Armed
	85	Kamov Ka-25Bsh 'Hormone'	Armed
	265	Kamov Ka-27 PL 'Helix A'	Armed
	148	Mil Mi-14PL/PW 'Haze A'	Armed
	6	HAL/Aerospatiale SA-316B Alouette III/Chetak	Armed
	59	Kamov Ka-25 'Hormone'	Armed
	81	Kamov Ka-25PS 'Hormone C'	Armed
	30	Kamov Ka-29TB 'Helix B'	Armed
	360	Mil Mi-6 'Hook A'	Armed
	90	Mil Mi-6 'Hook B'	Armed
	1490	Mil Mi-8 'Hip C'	Armed
	12	IAR (Kamov) Ka-126 Hoodlum-B	Utility
	44	Mil Mi-14BT 'Haze B'	Utility
	9	Mil Mi-24K 'Hind G2'	Utility
	110	Mil Mi-26 'Halo'	Utility
	3	Mil Mi-28 'Havoc'	Utility
	15	Mil Mi-4 'Hound'	Utility
	200	Mil Mi-8 'Hip J'	Utility
	10	Mil Mi-10 'Harke'	Unknown
	609	Mil Mi-2 'Hoplite'	Unknown
France			
	117	Aerospatiale SA-341F 'Canon'	Armed
	184	Eurocopter (Aerospatiale) SA-342M Gazelle	Armed
	26	Westland Lynx HAS. Mk2	Armed
	14	Westland Lynx HAS. Mk3	Armed
	3	Aerospatiale AS-332C Super Puma	Armed
	28	Aerospatiale AS-332M Super Puma	Armed
	35	Aerospatiale AS-350B Ecureuil	Armed
	52	Aerospatiale AS-355F1 Fennec	Armed
	6	Aerospatiale AS-365F Dauphin II	Armed
	20	Aerospatiale SA-321G/Ga Super Frelon	Armed
	160	Aerospatiale/Westland SA-330B Puma	Armed
	44	Eurocopter AS 555AN Fennec	Armed
	283	Aerospatiale Alouette II/Astazou (SE-3130) SA-318C	Utility
	2	Aerospatiale AS-332L Super Puma	Utility
	142	Aerospatiale SA-316 Alouette III (SA-319)	Utility
	5	Aerospatiale SA-365N Dauphin II (SA-360)	Utility
	18	Aerospatiale SA-565MA Dauphin Naval	Utility
	4	Eurocopter (Aerospatiale) AS-555UN Fennec	Utility
	36	Aerospatiale SA-313 Alouette II	Unknown
	35	Aerospatiale SA-341M Gazelle	Unknown
	1	Aerospatiale SA-361H Dauphin I	Unknown
Gabon			
	5	Aerospatiale SA-342L Gazelle	Armed
	1	Eurocopter (Aerospatiale) AS-532 Cougar	Armed
	3	Aerospatiale SA-316 Alouette III (SA-319)	Utility
	2	Eurocopter (Aerospatiale) AS-350 Ecureuil	Utility
	3	Aerospatiale SA-330C/H Puma	Unknown

Country	Number	Name	Role
Germany			
	10	PZL Swidnik (Mil) Mi-2RM/URP/US Hoplite	Armed
	51	Mil Mi-24 'Hind'	Armed
	14	Mil Mi-14PL/PW 'Haze A'	Armed
	19	Westland Sea Lynx Mk 88	Armed
	19	Westland Sea Lynx Mk 88	Armed
	3	Eurocopter (Aerospatiale) AS-332 Super Puma	Armed
	97	MBB VBH (Bo 105M)	Armed
	29	Mil Mi-8 'Hip C'	Armed
	10	Aerospatiale Alouette II/Astazou (SE-3130) SA-318C	Utility
	141	Aerospatiale SA-313B Alouette II	Utility
	185	Bell (204) UH-1D Iroquois	Utility
	184	Dornier-Bell UH-1D Iroquois	Utility
	2	Mil Mi-9 'Hip G'	Utility
	110	Sikorsky CH-53G Stallion (Sea Stallion)	Utility
	22	Westland Sea King Mk 41	Utility
	15	Mil Mi-8 'Hip P'	Unknown
	6	Mil Mi-8S 'Hip'	Unknown
	40	Mil Mi-8T 'Hip'	Unknown
Ghana			
	4	Aerospatiale Alouette II/Astazou (SE-3130) SA-318C	Utility
	2	Aerospatiale SA-316B Alouette III (SE-3160)	Utility
	2	Bell 212 (CH-135) Twin-Pac/Twin Huey	Utility
	2	Mil Mi-2 'Hoplite'	Unknown
Greece			
	17	Agusta-Bell 212ASW	Armed
	5	Sikorsky S-70B-6	Armed
	4	Aerospatiale SA-316 Alouette III (SA-319)	Utility
	1	Agusta A-109 Hirundo	Utility
	10	Agusta/Boeing CH-47C	Utility
	85	Bell (205) UH-1H Iroquois	Utility
	20	Bell 205A-1/UH-1H Iroquois	Utility
	1	Bell 212 (CH-135) Twin-Pac/Twin Huey	Utility
	30	Bredanardi/Schweizer-Hughes 300C	Utility
	14	AB-205A (Bell 205)	Unknown
	2	Agusta-Bell 206A	Unknown
	15	Bell 47G-3B-2A	Unknown
	3	Bell 47G-5	Unknown
Guatemala			
	6	Bell Military 412	Armed
	16	Bell (205) UH-1H Iroquois	Utility
	6	Bell 212 (CH-135) Twin-Pac/Twin Huey	Utility
	3	Sikorsky S-76A Plus	Utility
Guinea			
	1	Aerospatiale Gazelle (SA-342K)	Utility
	1	Aerospatiale SA-316B Alouette III (SE-3160)	Utility
	4	Mil Mi-4 'Hound'	Utility
	2	Aerospatiale/ICA-Brasov SA-330 Puma	Unknown
Guinea - Bissau			
	1	Aerospatiale Alouette II/Astazou (SE-3130) SA-318C	Utility
	2	Aerospatiale SA-316 Alouette III (SA-319)	Utility
Guyana			
	1	Bell Military 412	Armed
	3	Mil Mi-8 'Hip C'	Armed
	1	Bell 212 (CH-135) Twin-Pac/Twin Huey	Utility

Country	Number	Name	Role
Honduras			
	10	Bell 412SP	Armed
	15	Bell (205) UH-1H Iroquois	Utility
	1	Sikorsky S-76 Shadow	Utility
	7	TH-55 Hughes 269	Unknown
Hong Kong			
	8	Sikorsky S-76A Plus	Utility
Hungary			
	40	Mil Mi-24W 'Hind E'	Armed
	34	PZL Swidnik (Mil) Mi-2RM/URP/US Hoplite	Armed
	23	Mil Mi-17 'Hip'	Armed
	25	Mil Mi-8 'Hip C'	Armed
	7	Mil Mi-17P 'Hip K'	Utility
	1	Mil Mi-8 'Hip G'	Unknown
	65	Mil Mi-8TB/S 'Hip C'	Unknown
India			
	40	Mil Mi-25 'Hind' (export Mi-24D)	Armed
	13	Kamov Ka-28 'Helix A'	Armed
	3	Westland Sea King Mk 42A	Armed
	20	Westland Sea King Mk 42B	Armed
	6	Westland Sea King Mk 42C	Armed
	96	HAL/Aerospatiale SA-316B Alouette III/Chetak	Armed
	7	Kamov Ka-25PS 'Hormone C'	Armed
	50	Mil Mi-17 'Hip H'	Armed
	80	Mil Mi-8 'Hip C'	Armed
	40	HAL/Aerospatiale SA-315B Cheetah	Utility
	4	MD Hughes 300C	Utility
	2	Mil Mi-24 'Hind'	Utility
	10	Mil Mi-26 'Halo'	Utility
	30	Krishnar Mk II	Unknown
Indonesia			
	32	IPTN NBell 412	Armed
	9	Westland Wasp HAS-1	Armed
	4	MBB Bo-105 (NBo-105)	Armed
	61	Nurtanio/MBB NBo-105 BO-105	Armed
	3	Aerospatiale SA-316B Alouette III (SE-3160)	Utility
	2	Bell 204B Iroquois	Utility
	13	Bell 205A-1/UH-1H Iroquois	Utility
	20	MD Hughes 300C	Utility
	14	Nurtanio/Aerospatiale NSA-330 (SA-330L)	Utility
	12	Sikorsky UH-34T (S-58T)	Utility
	8	Aerospatiale/IPTN NAS-332B/L Super Puma	Unknown
	2	SA-313 (Alouette II)	Unknown
	8	Soloy-Bell 47G	Unknown

Country	Number	Name	Role
Iran			
	159	Bell AH-1J Cobra	Armed
	7	Agusta-Bell 212ASW	Armed
	68	Agusta/Boeing CH-47C	Utility
	250	Bell 214A	Utility
	5	Meridionali-Vertol CH-47C Chinook	Utility
	2	Sikorsky RH-53D Sea Stallion	Utility
	2	Sikorsky/Agusta AS-61A-4	Utility
	10	Sikorsky/Agusta SH-3D/TS	Utility
	35	Agusta-Bell 205A-1 Iroquois	Unknown
	50	Hkp-6 AB-206	Unknown
	39	Bell 214C	
Iraq			
	40	Mil Mi-24 'Hind'	Armed
	13	Aerospatiale SA-321GV Super Frelon	Armed
	20	Aerospatiale/Westland SA-330 Puma	Armed
	120	MBB Bo-105 (NBo-105)	Armed
	100	Mil Mi-17 'Hip'	Armed
	15	Mil Mi-6 'Hook'	Armed
	50	Aerospatiale Gazelle (SA-342K)	Utility
	3	Agusta A-109 Hirundo	Utility
	45	Bell 214ST SuperTransport	Utility
	30	MD Hughes 300C	Utility
	35	Aerospatiale SA-315C Alouette III	Unknown
	6	Agusta/Sikorsky AS-61TS	Unknown
Ireland			
	5	Aerospatiale AS-365F Dauphin II	Armed
	2	Aerospatiale SA-342L Gazelle	Armed
	8	Aerospatiale SA-316B Alouette III (SE-3160)	Utility
	2	Aerospatiale SA-365C Dauphin II (SA-360)	Utility
	5	Eurocopter AS-565MA Panther	Utility
Israel			
	6	Bell AH-1E HueyCobra	Armed
	42	Bell AH-1F Cobra	Armed
	40	Bell AH-1S Cobra	Armed
	18	McDonnell Douglas AH-64A Apache	Armed
	35	MD 500MD Scout Defender	Armed
	12	Bell (205) UH-1D Iroquois	Utility
	54	Bell 212 (CH-135) Twin-Pac/Twin Huey	Utility
	42	Sikorsky CH-53A Stallion	Utility
	22	Aerospatiale SA-366G (HH-65A) Dauphin	Unknown
	10	Agusta-Bell 205A-1 Iroquois	Unknown
	35	Bell OH-58A Kiowa	Unknown
	2	HH-65A	Unknown

Country	Number	Name	Role
Italy			
	6	Agusta A-129 Mangusta	Armed
	58	Agusta-Bell 212ASW	Armed
	60	Agusta-Bell 212ASW	Armed
	28	Agusta A.109EOA	Armed
	24	Agusta-Bell 412 Griffon	Armed
	36	Agusta/Sikorsky SH-3D Sea King	Armed
	50	Breda Nardi/MDH NH.500	Armed
	19	AB-47G	Utility
	22	Agusta A-109 Hirundo	Utility
	30	Agusta/Bell 204B Iroquois	Utility
	38	Agusta/Boeing CH-47C	Utility
	34	Sikorsky/Agusta HH-3F Pelican	Utility
	4	AB-205	Unknown
	92	AB-205A	Unknown
	2	S-61 SH-3	Unknown
	2	Sikorsky/Agusta AS-61A-4	Unknown
Ivory Coast			
	3	Aerospatiale/Westland SA-330 Puma	Armed
	2	Aerospatiale SA-316 Alouette III (SA-319)	Utility
	4	Aerospatiale SA-365C Dauphin II (SA-360)	Utility
Jamaica			
	4	Bell (205) UH-1H Iroquois	Utility
	2	Bell 212 (CH-135) Twin-Pac/Twin Huey	Utility
	1	Bell 222UT	Utility
Japan			
	63	Bell AH-1S Cobra/Sea Cobra	Armed
	83	Fuji/Bell AH-1F Cobra	Armed
	47	Mitsubishi (Sikorsky) SH-60J Sea Hawk	Armed
	2	Sikorsky (SH-60) S-70B-3 Seahawk	Armed
	3	Eurocopter (Aerospatiale) AS-332 Super Puma	Armed
	141	Fuji/Bell (HU-10) UH-1H Hiyodori (Iroquois)	Armed
	189	MD OH-6D/J Cayuse (Hughes 369)	Armed
	46	Mitsubishi (Sikorsky) UH-60J Night Hawk	Armed
	117	Mitsubishi/Sikorsky HSS-2A/B Chidori (Sea King)	Armed
	185	Mitsubishi/Sikorsky S-61A	Armed
	3	Aerospatiale AS-332L Super Puma	Utility
	32	Bell 212 (CH-135) Twin-Pac/Twin Huey	Utility
	54	Boeing CH-47D International Chinook	Utility
	61	Kawasaki/Boeing CH-47J Chinook	Utility
	66	Kawasaki/Vertol KV-107 (CH-46)	Utility
	34	Kawasaki/Vertol KV-107-II-4/4A Shirasagi (CH-46)	Utility
	12	MD OH-6D/J Cayuse (Hughes 369)	Utility
	21	Sikorsky S-70A-12 Black Hawk	Utility
	12	Sikorsky S-80M (CH-53)	Utility
	33	TH-55 Hughes 269	Utility
	1	Kawasaki/Bell 47G-2A Hibari	Unknown
	7	Kawasaki/Hughes OH-6J/D	Unknown
	2	MD Hughes 369	Unknown

14

Country	Number	Name	Role
Jordan			
	24	Bell AH-1F Cobra	Armed
	2	(Eurocopter) MBB Bo 105S	Armed
	12	Aerospatiale AS-332M Super Puma	Armed
	8	Eurocopter (Aerospatiale) AS-332 Super Puma	Armed
	8	MD Hughes 500D Defender	Armed
	3	Sikorsky S-70A-11 Black Hawk	Utility
	8	Sikorsky S-76 Shadow	Utility
	2	Sikorsky S-76B	Utility
	8	Aerospatiale SA-342 Gazelle	Unknown
Kenya			
	15	MD 500MD Scout Defender	Armed
	1	Aerospatiale SA-342 Gazelle	Armed
	3	Aerospatiale/Westland SA-330 Puma	Armed
	15	MD Hughes 500M Defender	Armed
	8	MD Hughes 500ME Defender	Armed
	10	Aerospatiale/IAR SA-330G Puma	Utility
Kuwait			
	4	Eurocopter (Aerospatiale) AS-532SC Cougar	Naval Attack
	16	Aerospatiale SA-342 Gazelle	Armed
	4	Eurocopter (Aerospatiale) AS-332 Super Puma	Armed
	9	Aerospatiale SA-330H Puma	Utility
Laos			
	2	Mil Mi-6 'Hook'	Armed
	10	Mil Mi-8 'Hip C'	Armed
Lebanon			
	4	Aerospatiale SA-342L Gazelle	Armed
	9	Aerospatiale/Westland SA-330L Puma	Armed
	2	Aerospatiale Alouette II/Astazou (SE-3130) SA-318C	Utility
	3	Aerospatiale Alouette III SA-319	Utility
	7	Aerospatiale SA-316B Alouette III (SE-3160)	Utility
	7	Agusta-Bell AB 212	Utility
Lesotho			
	3	Agusta-Bell 412 Griffon	Armed
	2	MBB Bo-105C (NBo-105)	Armed
	1	Soloy-Westland-Bell 47	Unknown
Libya			
	10	Mi-35 (Mi-25)	Armed
	14	Mil Mi-35 (Mi-25) 'Hind D/F'	Armed
	20	PZL Swidnik (Mil) Mi-2RM/URP/US Hoplite	Armed
	12	Mil Mi-14PL 'Haze A'	Armed
	25	Mil Mi-14PL/PW 'Haze A'	Armed
	12	Aerospatiale SA-321M Super Frelon	Armed
	5	Agusta-Bell 206	Armed
	10	Mil Mi-8 'Hip C'	Armed
	23	Aerospatiale SA-316 Alouette III (SA-319)	Utility
	2	Agusta A-109 Hirundo	Utility
	20	Agusta/Boeing CH-47C	Utility
	18	Boeing-Vertol CH-47C Chinook (CH-147)	Utility
	35	Mil Mi-24 'Hind'	Utility
	50	Mil Mi-4 'Hound'	Utility
	6	Aerospatiale SA-321GM Super Frelon	Unknown
Madagascar			
	6	Mil Mi-8 'Hip C'	Armed

14

Country	Number	Name	Role
Malawi			
	1	Aerospatiale AS-365F Dauphin II	Armed
	1	Aerospatiale SA-316B Alouette III (SE-3160)	Utility
	3	Aerospatiale SA-330J Puma	Utility
	4	Eurocopter (Aerospatiale) AS-350 Ecureuil	Utility
	1	Eurocopter (Aerospatiale) AS-355 Ecureuil II Twinstar	Utility
Malaysia			
	6	Westland Wasp HAS-1	Armed
	34	Sikorsky S-61A Nuri Upgrade	Armed
	32	Sikorsky S-61A-4 Nuri	Armed
	3	Nurtanio/Aerospatiale NAS-332M Super Puma	Utility
	6	Sikorsky S-76C	Utility
	25	Aerospatiale SA-316B Alouette III (SE-3160)	Utility
	7	Bell 47G (OH-13)	Utility
Mali			
	1	Mil Mi-8 'Hip C'	Armed
	2	Mil Mi-4 'Hound'	Utility
Malta			
	4	Agusta-Bell/Bell 47G-2	Unknown
Mexico			
	2	Aerospatiale Alouette III SA-319B Naval	Armed
	2	Aerospatiale/Westland SA-330 Puma	Armed
	2	Bell Military 412	Armed
	2	Eurocopter (Aerospatiale) AS-332 Super Puma	Armed
	12	Eurocopter (MBB) Bo-105CB	Armed
	1	Aerospatiale AS-332L Super Puma	Utility
	4	Aerospatiale SA-315 Lamma (Alouette II)	Utility
	1	Agusta A-109A	Utility
	8	Bell 205A Iroquois	Utility
	43	Bell 212 (CH-135) Twin-Pac/Twin Huey	Utility
	10	Sikorsky S-70A-24 Black Hawk (UH-60L)	Utility
	3	Bell 47	Unknown
Mongolia			
	10	Mil Mi-8 'Hip C'	Armed
	10	Mil Mi-24 'Hind'	Utility
	10	Mil Mi-4 'Hound'	Utility
Morocco			
	25	Aerospatiale SA-342K/L Gazelle	Armed
	2	Aerospatiale SA-315 Lamma (Alouette II)	Utility
	30	Aerospatiale SA-330F/G Puma	Utility
	2	Aerospatiale SA-365N Dauphin II (SA-360)	Utility
	30	Agusta-Bell 205A-1 Iroquois	Utility
	20	Agusta-Bell AB 212	Utility
	9	Agusta/Boeing CH-47C	Utility
	4	Aerospatiale SA-313 Alouette II	Unknown
Mozambique			
	12	Mil Mi-25 'Hind' (export Mi-24D)	Armed
	11	Mil Mi-8 'Hip C'	Armed
	6	Mil Mi-24 'Hind'	Utility
Nepal			
	2	Aerospatiale/Westland SA-330 Puma	Armed
	2	Eurocopter (Aerospatiale) AS-332 Super Puma	Armed
	2	Aerospatiale SA-330C/G Puma	Utility
	3	HAL/Aerospatiale/IAR Alouette III/Chetak	Utility

Country	Number	Name	Role
Netherlands			
	64	Aerospatiale Alouette III SA-319B Naval	Armed
	29	MBB Bo-105CB/DB (NBo-105)	Armed
	6	Westland Lynx HAS. Mk25	Armed
	10	Westland Lynx HAS. Mk27	Armed
	8	Westland Lynx HAS. Mk81	Armed
	22	Westland Lynx UH-14ASH-14B/SA-14C	Armed
New Zealand			
	9	Westland Wasp HAS-1	Armed
	14	Bell (205) UH-1H Iroquois	Utility
	4	Bell 47G-3B-1/2	Utility
Nicaragua			
	33	Mil Mi-8/-17 'Hip E/H'	Armed
	2	Aerospatiale SA-316 Alouette III (SA-319)	Utility
	1	Hughes OH-6A Cayuse	Utility
	5	Mil Mi-2 'Hoplite'	Utility
	14	Mil Mi-24 'Hind'	Utility
Nigeria			
	6	Aerospatiale/Westland SA-330L Puma	Armed
	2	Bell 412SP	Armed
	2	Eurocopter (Aerospatiale) AS-332 Super Puma	Armed
	24	MBB Bo-105CB/D (NBo-105)	Armed
	14	MD Hughes 300C	Utility
	3	Westland Lynx Mk 89	Utility
	2	Bell 220	Unknown
North Korea			
	100	PZL Swidnik (Mil) Mi-2RM/URP/US Hoplite	Armed
	30	Harbin Z-5 (Mi-4)	Armed
	80	MD Hughes 500D Defender	Armed
	50	Mil Mi-24 'Hind'	Utility
	5	Mil Mi-24 'Hind'	Utility
	20	Pinkiang H-4 (Mil Mi-4) 'Hound'	Utility
	1	Schweizer-Hughes 300C	Utility
	70	Mil Mi-8/-17 'Hip'	Armed
Norway			
	17	Bell Military 412	Armed
	29	Bell UH-1B/C Iroquois	Armed
	10	Westland Sea King Mk 43/43A	Armed
	1	Westland Sea King Mk 43A	Armed
Oman			
	2	Eurocopter (Aerospatiale) AS-332 Super Puma	Armed
	20	Agusta-Bell 205A-1 Iroquois	Utility
	10	Bell 214B Big Lifter	Utility
	3	Fuji/Bell HU-18 AB-212	Utility

Country	Number	Name	Role
Pakistan			
	30	Bell AH-1F Cobra	Armed
	3	Kaman SH-2F Seasprite	Armed
	4	Aerospatiale SA-321 Super Frelon	Armed
	24	Aerospatiale/Dhamial/IAR SA-316B Alouette III	Armed
	16	Mil Mi-8 'Hip'	Armed
	6	Westland Sea King Mk 45	Armed
	6	Aerospatiale Alouette III SA-319	Utility
	23	Aerospatiale/Dhamial SA-316B Alouette III	Utility
	7	Agusta-Bell 205A-1 Iroquois	Utility
	5	Bell (205) UH-1H Iroquois	Utility
	35	IAR (Eurocopter France) IAR-330L Puma	Utility
	6	IAR/Aerospatiale IAR-315 (SA-315B)	Utility
Panama			
	1	Eurocopter (Aerospatiale) AS-332 Super Puma	Armed
	2	Aerospatiale AS-332L Super Puma	Utility
	2	Bell 205 (CH-118) Iroquois	Utility
	3	Bell 212 (CH-135) Twin-Pac/Twin Huey	Utility
	10	Bell UH-1H/N Iroquois	Utility
	1	Bell UH-1N (Bell 212)	Utility
	1	Bell UH-1B Iroquois	Unknown
Papua New Guinea			
	4	Mil Mi-17 'Hip H'	Armed
	4	Bell (205) UH-1H Iroquois	Utility
	3	Arava	Unknown
Paraguay			
	2	Bell UH-1B Iroquois	Armed
	2	Bell 47G (OH-13)	Utility
	3	Helibras/Aerospatiale HB-350[CH-50] Esquilo (AS-350)	Utility
	4	Hiller UH-12 (OH-23)	Utility
	2	Hiller UH-12E (OH-23)	Utility
	1	MD Hughes 300C	Utility
Peru			
	12	Mil Mi-25 'Hind' (export Mi-24D)	Armed
	5	Agusta-Bell 212ASW	Armed
	3	Aerospatiale AS-350B1 Ecureuil	Armed
	6	Agusta/Sikorsky SH-3D Sea King	Armed
	3	Bell Military 412	Armed
	10	MBB Bo-105CBS (NBo-105)	Armed
	23	Mil Mi-17 'Hip H'	Armed
	2	Mil Mi-6 'Hook'	Armed
	8	Mil Mi-6 'Hook'	Armed
	30	Mil Mi-8 'Hip C'	Armed
	6	Mil Mi-8 'Hip C'	Armed
	3	Aerospatiale Alouette II/Astazou (SE-3130) SA-318C	Utility
	6	Aerospatiale SA-315 Lamma (Alouette II)	Utility
	10	Aerospatiale SA-316 Alouette III (SA-319)	Utility
	2	Aerospatiale SA-316 Alouette III (SA-319)	Utility
	12	Bell (205) UH-1H Iroquois	Utility
	15	Bell 212 (CH-135) Twin-Pac/Twin Huey	Utility
	5	Bell 214ST SuperTransport	Utility
	12	Bell 47G (OH-13)	Utility
	10	Enstrom F28F Falcon	Utility
	12	Bell 47G-3B/-5A Sioux	Unknown

Country	Number	Name	Role
Philippines			
	2	Aerospatiale/Westland SA-330 Puma	Armed
	28	MD 500MD Defender II	Armed
	14	PADC/MBB Bo-105C	Armed
	55	Bell 205 (CH-118) Iroquois	Utility
	2	Bell 212 (CH-135) Twin-Pac/Twin Huey	Utility
	2	Bell 214B Big Lifter	Utility
	21	MD 500MG Night Fox	Utility
	2	Sikorsky S-70A-5 Black Hawk	Utility
	75	Bell UH-1B/H/205A-1 Iroquois	Unknown
	14	Sikorsky AUH-76/S-76	Unknown
Poland			
	30	Mil Mi-24W 'Hind E'	Armed
	18	Mil MI-14PW 'Haze'	Armed
	8	Kamov Ka-26 'Hoodlum'	Armed
	3	Mil Mi-17 'Hip H'	Armed
	3	Mil Mi-6 'Hook'	Armed
	5	Mil Mi-8 'Hip S'	Armed
	50	Mil Mi-8TB 'Hip C'	Armed
	3	PZL Swidnik W-3 Sokol (Falcon)	Armed
	5	Mil Mi-14PS 'Haze C'	Utility
	130	Mil Mi-2 'Hoplite'	Utility
	21	Mil Mi-28 'Havoc'	Utility
Portugal			
	5	Westland Super Lynx Mk 95	Armed
	7	Aerospatiale Alouette II/Astazou (SE-3130) SA-318C	Utility
	35	Aerospatiale SA-316B Alouette III (SE-3160)	Utility
	10	Aerospatiale SA-330C/L/S Puma	Utility
	52	Bell (205) UH-1H Iroquois	Utility
Qatar			
	2	Aerospatiale SA-341G Gazelle	Armed
	12	Aerospatiale SA-342L Gazelle	Armed
	3	Westland Lynx HAS. Mk28	Armed
	12	Westland Commando Sea King HC.Mk4	Utility
Romania			
	230	ICA-Brasov/Aerospatiale IAR-316B Alouette III	Armed
	6	Mil Mi-17 'Hip'	Armed
	28	Mil Mi-8 'Hip C'	Armed
	4	Aerospatiale SA-365N Dauphin II (SA-360)	Utility
	100	IAR (Eurocopter France) IAR-330L Puma	Utility
Rwanda			
	6	Aerospatiale SA-342L Gazelle	Armed
	7	Aerospatiale SA-316B Alouette III (SE-3160)	Utility

Country	Number	Name	Role
Saudi Arabia			
	12	McDonnell Douglas AH-64A Apache	Armed
	12	Eurocopter (Aerospatiale) AS-532SC Cougar	Armed
	12	Aerospatiale AS-332B Super Puma	Armed
	6	Aerospatiale AS-532UC Super Puma	Armed
	15	Bell 406CS (MH-58D) Combat Scout (export)	Armed
	20	Eurocopter (Aerospatiale) AS-565SA Panther	Armed
	30	Aerospatiale SA-365N Dauphin II (SA-360)	Utility
	14	Agusta-Bell 205A-1 Iroquois	Utility
	27	Bell 212 (CH-135) Twin-Pac/Twin Huey	Utility
	4	Eurocopter AS-565MA Panther	Utility
	18	Kawasaki/Vertol KV-107-II (CH-46)	Utility
	8	Sikorsky (UH-60L) S-70B Black Hawk (SH-60)	Utility
	13	Sikorsky S-70A-1 Desert Hawk	Utility
	16	Sikorsky S-70A-1L Desert Hawk	Utility
	4	Aerospatiale SA-565SC Panther	Unknown
	1	Agusta-Bell 204C Iroquois	Unknown
Senegambia			
	1	Aerospatiale SA-341H Gazelle	Armed
	2	Aerospatiale Alouette II/Astazou (SE-3130) SA-318C	Utility
	2	Aerospatiale SA-330F Puma	Utility
Seychelles			
	2	HAL/Aerospatiale SA-316B Alouette III/Chetak	Armed
Sierra Leone			
	2	Aerospatiale AS-355F2 Fennec	Armed
	1	MBB Bo-105 (NBo-105)	Armed
Singapore			
	27	Aerospatiale AS-332M Super Puma	Armed
	6	Aerospatiale SA-350E Ecureuil	Armed
	18	Bell UH-1B Iroquois	Armed
	22	Eurocopter (Aerospatiale) AS-332 Super Puma	Armed
	16	Bell (205) UH-1H Iroquois	Utility
	5	Bell 205 (CH-118) Iroquois	Utility
	6	Bell 205A Iroquois	Utility
	6	Eurocopter (Aerospatiale) AS-550 Fennec	Unknown
Somalia			
	2	Mil Mi-8 'Hip C'	Armed
	4	Bell 212 (CH-135) Twin-Pac/Twin Huey	Utility
	5	Mil Mi-4 'Hound'	Utility
South Africa	8	Westland Wasp HAS-1	Armed
	1	MBB Bo-105 (NBo-105)	Armed
	30	Aerospatiale Alouette II/Astazou (SE-3130) SA-318C	Utility
	68	Aerospatiale SA-316B Alouette III (SE-3160)	Utility
	65	Aerospatiale SA-330F/J/L Puma	Utility
	1	Aerospatiale SA-365C Dauphin II (SA-360)	Utility

Country	Number	Name	Role
South Korea			
	62	Bell AH-1F Cobra	Armed
	187	MD 500MD Scout Defender/TOW Defender	Armed
	25	MD Hughes/KAL 500MD	Armed
	3	Bell Military 412	Armed
	3	Eurocopter (Aerospatiale) AS-332 Super Puma	Armed
	12	Westland Super Lynx Mk 99	Armed
	2	Aerospatiale AS-332L Super Puma	Utility
	10	Aerospatiale SA-316 Alouette III (SA-319)	Utility
	130	Bell (205) UH-1H Iroquois	Utility
	5	Bell (205) UH-1H Iroquois	Utility
	7	Bell 212 (CH-135) Twin-Pac/Twin Huey	Utility
	75	Bell UH-1B/H Iroquois	Utility
	24	Boeing CH-47D International Chinook	Utility
	20	Boeing-Vertol CH-47D Chinook (CH-147)	Utility
	175	MD Hughes 500 Defender	Utility
	5	Sikorsky UH-60P (S. Korean UH-60L)	Utility
	30	UH-23	Utility
	3	Kawasaki/Bell KH-4 (Bell 47)	Unknown
Spain	16	Agusta-Bell 212ASW	Armed
	6	Sikorsky (SH-60) S-70B Seahawk	Armed
	9	Sikorsky SH-3H Sea King	Armed
	12	Sikorsky SH-60B Seahawk	Armed
	4	Aerospatiale Alouette III SA-319B Naval	Armed
	31	Aerospatiale AS-332B Super Puma	Armed
	30	Eurocopter (Aerospatiale) AS-332 Super Puma	Armed
	60	Eurocopter Bo-105 (NBo-105)	Armed
	9	MBB Bo-105C	Armed
	3	Westland Commando Sea King SH-3D	Armed
	5	Aerospatiale SA-330H/J Puma	Utility
	66	Agusta-Bell 205A/UH-1H Iroquois	Utility
	50	Bell (205) UH-1H Iroquois	Utility
	6	Bell 212 (CH-135) Twin-Pac/Twin Huey	Utility
	20	Bell 47G (OH-13)	Utility
	17	Bell OH-58A Kiowa	Utility
	9	Bell UH-1C Iroquois	Utility
	9	Boeing CH-47D International Chinook	Utility
	19	Boeing HT.17 Chinook (Spa. vers.)	Utility
	17	Hughes 300C	Utility
	10	Hughes 500M Defender	Utility
	10	Sikorsky S-76C	Utility
	3	Bell UH-1B Iroquois	Unknown
	68	CASA-MBB Bo-105ATH	Unknown
	18	CASA-MBB Bo-105GSH	Unknown
	13	CASA-MBB Bo-105LOH	Unknown
Sri Lanka			
	4	Bell Military 412	Armed
	2	Aerospatiale SA-365C Dauphin II (SA-360)	Utility
	13	Bell 212 (CH-135) Twin-Pac/Twin Huey	Utility

Country	Number	Name	Role
Sudan			
	11	Agusta-Bell 412 Griffon	Armed
	12	MBB Bo-105 (NBo-105)	Armed
	14	Mil Mi-8 'Hip C'	Armed
	15	Aerospatiale SA-316 Alouette III (SA-319)	Utility
	11	Bell 212 (CH-135) Twin-Pac/Twin Huey	Utility
	14	IAR (Eurocopter France) IAR-330L Puma	Utility
	2	Mil Mi-24 'Hind'	Utility
	4	Mil Mi-4 'Hound'	Utility
Suriname			
	2	Aerospatiale SA-316 Alouette III (SA-319)	Utility
	1	Bell 205 (CH-118) Iroquois	Utility
Sweden			
	14	Boeing-Vertol/Kawasaki Hkp-4B/4C (KV-107-5)	Armed
	10	Aerospatiale Hkp-10 (AS-332M1) Super Puma	Armed
	10	Eurocopter (Aerospatiale) AS-332 Super Puma	Armed
	20	Eurocopter Bo-105 (NBo-105)	Armed
	1	Aerospatiale Hkp-2 (Aloutte II/SE-3130)	Utility
	18	Agusta-Bell Hkp-3B (AB 204B) Iroquois	Utility
	19	Agusta-Bell Hkp-6A (AB-206A) JetRanger	Utility
	29	Agusta-Bell Hkp-6B (AB-206A) JetRanger	Utility
	4	Boeing-Vertol/Kawasaki Hkp-4A (KV-107-II-4)	Utility
	4	Eurocopter (MBB) Bo-105CBS	Utility
	26	Hughes Hkp-5B (Hughes 300C)	Utility
Switzerland			
	9	Aerospatiale AS-332M1 Super Puma	Armed
	15	Eurocopter (Aerospatiale) AS-532 Cougar	Armed
	21	Aerospatiale Alouette II/Astazou (SE-3130) SA-318C	Utility
	78	Aerospatiale SA-316B Alouette III (SE-3160)	Utility
Syria			
	50	Mil Mi-25 'Hind' (export Mi-24D)	Armed
	20	Mil Mi-14 'Haze'	Armed
	12	Mil Mi-14PL 'Haze A'	Armed
	50	Aerospatiale SA-342L Gazelle	Armed
	10	Kamov Ka-25PS 'Hormone C'	Armed
	40	Mil Mi-17 'Hip'	Armed
	10	Mil Mi-6 'Hook'	Armed
	130	Mil Mi-8 'Hip C'	Armed
	20	Mil Mi-2 'Hoplite'	Utility
Taiwan			
	12	MD 500MD Defender II	Armed
	100	AIDC/Bell UH-1H Iroquois	Utility
	12	Bell 47G (OH-13)	Utility
	6	Boeing CH-47C Chinook (CH-147)	Utility
	6	Hughes OH-6A Cayuse	Utility
	6	MD Hughes 500 Defender	Utility
	5	Sikorsky CH-34 Choctaw	Utility
	1	Sikorsky S-62A	Utility
	24	Sikorsky S-70C Black Hawk/Seahawk	Utility
	10	Sikorsky S-70C(M)-1 Thunderhawk (SH-60B)	Utility
	14	Sikorsky S-70C-2	Utility
	24	Kawasaki/Bell KH-4 (Bell 47)	Unknown
Tanzania			
	4	Agusta-Bell 205A Iroquois	Utility

Country	Number	Name	Role
Thailand			
	8	Bell AH-1F Cobra	Armed
	45	Agusta-Bell 212ASW	Armed
	101	Bell UH-1H/A/B Iroquois	Armed
	6	Kaman SH-2 Sea Sprite	Armed
	3	Bell 214B Big Lifter	Utility
	8	Bell 214ST SuperTransport	Utility
	12	Bell 47G/OH-13H Trooper	Utility
	12	Boeing CH-47D International Chinook	Utility
	17	Hughes TH-55A Osage	Utility
	2	Kawasaki/Vertol KV-107-II (CH-46)	Utility
	48	Schweizer Model TH-300C	Utility
	18	Sikorsky UH-34T (S-58T)	Utility
	18	Thai-Am/Sikorsky S-58T Twin Pac	Unknown
Togo			
	1	Aerospatiale/Westland SA-330 Puma	Armed
	1	Eurocopter (Aerospatiale) AS-332 Super Puma	Armed
	2	Aerospatiale SA-313B Alouette II	Utility
	2	Aerospatiale SA-315B Lama	Utility
Tunisia			
	1	Aerospatiale AS-365F Dauphin II	Armed
	1	Aerospatiale/Westland SA-330 Puma	Armed
	21	Agusta-Bell 205	Armed
	5	Agusta-Bell 205A-1 Iroquois	Armed
	5	Aerospatiale Alouette II/Astazou (SE-3130) SA-318C	Utility
	6	Aerospatiale SA-313 Alouette II	Utility
	4	Aerospatiale SA-316B Alouette III (SE-3160)	Utility
	5	Aerospatiale SA-341 Gazelle	Utility
	6	Aerospatiale SA-350 Ecureuil	Utility
	4	Bell (205) UH-1H Iroquois	Utility
	6	Bell UH-1N (Bell 212)	Utility
Turkey			
	9	Bell AH-1W SuperCobra Viper	Armed
	15	Agusta-Bell 212ASW	Armed
	5	Aerospatiale AS-532 Cougar	Armed
	3	Agusta-Bell 204	Armed
	20	Bell UH-1B Iroquois	Armed
	60	Aerospatiale SA-313B Alouette II	Utility
	20	Agusta-Bell 204AS	Utility
	130	Agusta-Bell 205A-1 Iroquois	Utility
	4	AS-330	Utility
	30	Bell (205) UH-1D Iroquois	Utility
	70	Bell (205) UH-1H Iroquois	Utility
	30	Bell 47G/OH-13H Trooper	Utility
	30	Hughes 300C	Utility
	25	Bell TH-13T Sioux	Unknown
	70	TH-55 Hughes 269	Unknown
Uganda			
	11	Agusta-Bell 412 Griffon	Armed
	4	Mil Mi-17 'Hip'	Armed
	2	Mil Mi-8 'Hip C'	Armed
	3	Bell 205A-1/UH-1H Iroquois	Utility
	1	Bell 212 (CH-135) Twin-Pac/Twin Huey	Utility
	3	Bell 214	Utility
	1	Westland-Bell 47G-3B-1 Sioux	Unknown

Country	Number	Name	Role
United Arab Emirates			
	7	Aerospatiale Alouette III SA-319B Naval	Armed
	6	Aerospatiale AS-332F Super Puma Naval	Armed
	1	Aerospatiale AS-350B Ecureuil	Armed
	11	Aerospatiale SA-342L Gazelle	Armed
	3	Agusta-Bell 412 Griffon	Armed
	3	MBB Bo-105 (NBo-105)	Armed
	10	Aerospatiale AS-532UC/UL/UL1 Super Puma	Utility
	11	Aerospatiale SA-330C/F Puma	Utility
	3	Bell 205A-1/UH-1H Iroquois	Utility
	4	Bell 214B Big Lifter	Utility
	1	Sikorsky S-76 Shadow	Utility
United Kingdom			
	31	Westland Lynx HAS. Mk3	Armed
	61	Aerospatiale SA-342L Gazelle	Armed
	44	Aerospatiale/Westland SA-330L Puma	Armed
	82	Westland Lynx HAS-3/3GM/3S/3CTS	Armed
	70	Westland Scout AH.1	Armed
	56	Westland Sea King HAS. Mk 1	Armed
	81	Westland Sea King HAS. Mk 2	Armed
	10	Westland Sea King HAS. Mk 2A	Armed
	144	Westland Sea King HAS. Mk 5	Armed
	35	Westland Sea King HAS. Mk 6	Armed
	37	Westland Sea King HAS. Mk 6	Armed
	179	Westland-Aerospatiale SA-341B Gazelle AH.1	Armed
	75	Westland/Agusta EH-101 Merlin	Armed
	33	Aerospatiale SA-341 Gazelle	Utility
	4	Agusta A-109 Hirundo	Utility
	34	Boeing Chinook HC.Mk 1/2 (RAF vers.) CH47-352	Utility
	11	Sea King AEW-2	Utility
	7	Sikorsky S-76C	Utility
	38	Westland Commando Sea King HC.Mk4	Utility
	113	Westland Lynx AH. Mk1	Utility
	3	Westland Lynx AH. Mk5	Utility
	11	Westland Lynx AH. Mk7	Utility
	16	Westland Lynx AH. Mk9	Utility
	44	Westland Sea King HAR. Mk 3	Utility
	5	Westland Sea King HAR. Mk 5	Utility
	64	Westland Wessex HC.2/HCC.4/HC.5C/HU.5	Utility
	2	Westland/Sikorsky WS-70L Black Hawk (S-70C)	Utility
	1	Bell 222UT	Unknown

Country	Number	Name	Role
United States	90	Bell AH-1E HueyCobra	Armed
	520	Bell AH-1F Cobra	Armed
	150	Bell AH-1G Cobra	Armed
	120	Bell AH-1J Sea Cobra	Armed
	90	Bell AH-1P Cobra	Armed
	160	Bell AH-1S Cobra	Armed
	105	Bell AH-1W SuperCobra Viper	Armed
	807	McDonnell Douglas AH-64A Apache	Armed
	137	SH-60B Sea Hawk (UH-60)	Armed
	60	SH-60F Sea Hawk (UH-60)	Armed
	15	Bell OH-58D (Bell 406) Kiowa Prime Chance	Armed
	1284	Bell OH-58D (Bell 406) Kiowa Warrior	Armed
	18	Bell OH-58D (Bell 406) Stealth Kiowa Warrior	Armed
	205	Boeing-Vertol HH+UH-46 (CH-46E) Sea Knight	Armed
	40	Boeing-Vertol HH+UH-46 (CH-46F) Sea Knight	Armed
	95	Boeing-Vertol HH+UH-46D (CH-46) Sea Knight	Armed
	6	HH-1K	Armed
	98	Kaman SH-2F/G Sea Sprite	Armed
	148	Kaman SH-2F/G Sea Sprite/Super Sea Sprite	Armed
	4	MD EH-6B (Hughes 500/530 Defender)	Armed
	580	OH-58C Kiowa (Bell 206)	Armed
	6	Sikorsky HH-3A Sea King	Armed
	323	Sikorsky S-80/H-53E (CH-53E Super Stallion/MH-53E)	Armed
	120	Sikorsky SH-3 (Sea King)	Armed
	70	Bell (205) UH-1H Iroquois	Utility
	30	Bell EH-1H/EH-1X	Utility
	20	Bell TAH-1S Cobra	Utility
	3215	Bell UH-1B/C/D/H/M Iroquois (Bell 204/205)	Utility
	120	Bell UH-1N (Bell 212)	Utility
	4	Bell VH-1N Iroquois	Utility
	210	Boeing-Vertol CH-46D/E Sea Knight (CH-113)	Utility
	155	Boeing-Vertol CH-47C Chinook (CH-147)	Utility
	300	Boeing-Vertol CH-47D Chinook (CH-147)	Utility
	17	Boeing-Vertol MH-47E	Utility
	6	HH-53 (CH-53)	Utility
	9	Hughes AH-6C	Utility
	6	Hughes AH-6F	Utility
	6	Hughes AH-6G	Utility
	2	Hughes EH-6E Defender	Utility
	4	Hughes MH-6B	Utility
	5	Hughes MH-6E	Utility
	10	Hughes MH-6F	Utility
	280	Hughes OH-6A Cayuse	Utility
	31	MH-53E	Utility
	41	MH-53J Super Jolly	Utility
	18	RH-53 (CH-53)	Utility
	30	Schweizer Model TH-300C	Utility
	18	Sikorsky (HH-60H) S-70B Jayhawk	Utility
	20	Sikorsky (HH-60J) S-70B Jayhawk	Utility
	55	Sikorsky CH-3E/HH-3E Jolly Green Giant	Utility
	46	Sikorsky CH-54A Tarhe	Utility
	25	Sikorsky CH-54B Tarhe	Utility
	66	Sikorsky EH-60C (UH-60) Black Hawk	Utility
	56	Sikorsky MH-53E Sea Dragon	Utility
	60	Sikorsky MH-60G Pave Hawk	Utility

Country	Number	Name	Role
United States (continued)			
	24	Sikorsky RH-53D Sea Stallion	Utility
	4	Sikorsky TH-53A Super Jolly	Utility
	1041	Sikorsky UH-60A Black Hawk (SH-60)	Utility
	11	Sikorsky VH-3D Sea King	Utility
	9	Sikorsky VH-60 Desert Hawk	Utility
	36	Sikorsky/Agusta HH-3F Pelican	Utility
	9	VH-60A	Utility
	96	HH-65A	Unknown
Uruguay			
	3	Bell (205) UH-1H Iroquois	Utility
	2	Bell 212 (CH-135) Twin-Pac/Twin Huey	Utility
	1	Bell 47G/OH-13H Trooper	Utility
	2	Bell UH-1B Iroquois	Utility
	1	Sikorsky CH-34 Choctaw (HH-34)	Utility
	1	Sikorsky UH-34T (S-58T)	Utility
	1	Bell 222	Unknown
Vietnam			
	30	Mil Mi-24 'Hind D'	Armed
	8	Kamov Ka-25Bsh 'Hormone'	Armed
	17	Kamov Ka-25 'Hormone'	Armed
	10	Mil Mi-6 'Hook'	Armed
	60	Mil Mi-8 'Hip C'	Armed
	50	Bell UH-1N (Bell 212)	Utility
	10	Boeing-Vertol CH-47 Chinook (CH-147)	Utility
Venezuela			
	14	Agusta-Bell 212ASW	Armed
	8	Aerospatiale AS-332B Super Puma	Armed
	6	Agusta-Bell 412 Griffon	Armed
	26	Bell UH-1B/D/H Iroquois	Armed
	8	Eurocopter (Aerospatiale) AS-332 Super Puma	Armed
	12	Aerospatiale SA-316B Alouette III (SE-3160)	Utility
	6	Agusta A-109 Hirundo	Utility
	4	Agusta A-109 Hirundo	Utility
	4	Agusta-Sikorsky AS-61R (HH-3F) Pelican	Utility
	3	Bell 214ST SuperTransport	Utility
	1	Bell 47G (OH-13)	Utility
	2	Bell UH-1N (Bell 212)	Utility
	2	UH-13 (Bell 47J)	Unknown
Yemen			
	20	Mil Mi-24 'Hind D'	Armed
	40	Mil Mi-8 'Hip C'	Armed
	2	Agusta-Bell 214	Utility
	5	Agusta-Bell AB 212	Utility
	2	Agusta/Bell 204B Iroquois	Utility
	2	Mil Mi-4 'Hound'	Utility

Country	Number	Name	Role
Yugoslavia (former)			
	10	Kamov Ka-25Bsh 'Hormone'	Armed
	2	Kamov Ka-28 'Helix A'	Armed
	4	Mil Mi-14PL 'Haze A'	Armed
	220	Aerosptiale/SOKO SA-341H/342L Partizan (Gazelle)	Armed
	15	ICA-Brasov/Aerospatiale IAR-316B Alouette III	Armed
	110	Mil Mi-8 'Hip C'	Armed
	2	Agusta A-109 Hirundo	Utility
	5	Agusta-Bell 205A-1 Iroquois	Utility
Zaire			
	1	Eurocopter (Aerospatiale) AS-332 Super Puma	Armed
	1	Aerospatiale AS-332L Super Puma	Utility
	7	Aerospatiale SA-316 Alouette III (SA-319)	Utility
	9	Aerospatiale SA-330C Puma	Utility
	6	Bell 47G (OH-13)	Utility
Zambia			
	12	Mil Mi-8 'Hip C'	Armed
	4	Agusta-Bell 205A Iroquois	Utility
	5	Agusta-Bell AB 212	Utility
	12	Bell 47G (OH-13)	Utility
Zimbabwe			
	11	Agusta-Bell 412 Griffon	Armed
	20	Aerospatiale SA-316 Alouette III (SA-319)	Utility
	5	Agusta-Bell 205A Iroquois	Utility

2. Surface-to-Air Missile Data

Country	Quantity (1991)	Surface-to-Air Missile
Afghanistan		
	12	SA-3 launchers
	16	SA-13 launchers
	18	SA-2 launchers
Albania		
	24	SA-2 launchers
Algeria		
	40	SA-9 launchers
	32	SA-13 launchers
	44	SA-3 launchers
	60	SA-6 launchers
	20	SA-8 launchers
	42	SA-2 launchers
Angola		
	30	SA-9 launchers
	30	SA-13 launchers
	48	SA-3 launchers
	72	SA-6 launchers
	72	SA-8 launchers
	18	SA-2 launchers
Argentina		
	6	Grenadier (Modified Roland 2)
	4	Roland 2 (Shelter mounted)
	10	Tigercat launchers
Australia		
	20	Rapier launchers
Belgium		
Benin		
	4	SA-9 launchers
Brazil		
	4	Roland launchers
Brunei		
	12	Rapier launchers
Bulgaria		
	50	SA-9 launchers
	20	SA-13 launchers
	136	SA-3 launchers
	40	SA-6 launchers
	132	SA-2 launchers
	27	SA-4 launchers
Canada		
	36	ADATS (STA-AA) launchers
Chile		
	12	Crotale/Cactus fire units

Country	Quantity (1991)	Surface-to-Air Missile
Cuba		
	60	SA-9 launchers
	40	SA-13 launchers
	48	SA-3 launchers
	4	SA-6 launchers
	144	SA-2 launchers
Czechoslovakia		
	80	SA-9 launchers
	100	SA-13 launchers
	120	SA-3 launchers
	120	SA-6 launchers
	40	SA-8 launchers
	18	SA-10b
	120	SA-2 launchers
	27	SA-4 launchers
	18	SA-5 launchers
Denmark		
		Man-Portable
	48	I-Hawk launchers
Egypt		
	20	SA-9 launchers
	48	Crotale launcher
	26	Chaparral launcher
	240	SA-3 launchers
	75	SA-6 launcher
	78	Hawk launchers
	400	SA-2 launchers
Ethiopia		
	32	SA-3 launchers
	36	SA-2 launchers
Finland		
	20	VT-1 Crotale launcher
	48	SA-3 launchers
Former Soviet Union		
	450	SA-9 launchers
	250	SA-11 launchers
	1200	SA-13 launchers
	200	SA-15 launchers
	1250	SA-3 launchers
	850	SA-6 launchers
	1000	SA-8 launchers
	1700	SA-10A launchers
	1500	SA-10B launchers
	45	SA-12A launchers
	3200	SA-1 launchers
	2730	SA-2 launchers
	1300	SA-4 launchers
	2000	SA-5 launchers
France		
	48	Crotale (Firing Unit)
	190	Roland II launchers
	69	Hawk launchers

Country	Quantity (1991)	Surface-to-Air Missile
Germany		
	60	SA-9 launchers
	40	SA-13 launchers
	95	Roland launchers
	120	SA-6 launchers
	40	SA-8 launchers
	48	SA-3 launchers
	216	Hawk launchers
	132	SA-2 launchers
	54	SA-4 launchers
	320	Patriot launchers
	216	Nike Hercules launchers
	12	SA-5 launchers
Greece		
	36	Nike Hercules launchers
	42	Hawk launchers
Guinea		
	16	SA-9 launchers
	4	SA-6 launchers
Hungary		
	44	SA-9 launchers
	12	SA-13 launchers
	24	SA-3 launchers
	44	SA-6 launchers
	8	SA-8 launchers
	120	SA-2 launchers
	27	SA-4 launchers
India		
	200	SA-9 launchers
	40	Tigercat launchers
	48	SA-3 launchers
	185	SA-6 launchers
	48	SA-8 launchers
	150	SA-2 launchers
Iran		
	45	Rapier launchers
	15	Tigercat launchers
	222	Hawk launchers
	60	HQ-2 launchers (SA-2)
Iraq		
	100	SA-9 launchers
	60	SA-13 launchers
	27	Roland launchers
	100	SA-3 launchers
	180	SA-6 launchers
	50	SA-8 launchers
	48	Hawk launchers
	120	SA-2 launchers
Israel		
	52	Chaparral launchers
	1	SA-6 (tech eval)
	68	Hawk launchers
	16	Patriot launchers

Country	Quantity (1991)	Surface-to-Air Missile
Italy		
	48	Spada/Aspide
	60	Hawk launchers
	20	Patriot launchers
	96	Nike Hercules launchers
Japan		
	10	Tan-Sam launchers
	192	Hawk launchers
	180	Nike Hercules launchers
Jordan		
	40	SA-9 launchers
	40	SA-13 launchers
	20	Roland launchers
	16	SA-6 launchers
	50	SA-8 launchers
	56	Hawk launchers
Kenya		
	25	Mistral launchers
	5	Tigercat launchers
Kuwait		
	20	SA-8 launchers
	27	Hawk launchers
	6	Patriot launchers
Libya		
	60	SA-9 launchers
	60+	SA-13
	30	Crotale launchers
	132	SA-3 launchers
	160	SA-6
	20	SA-8 launchers
	108	SA-2 launchers
Madagascar		
	4	SA-9 launchers
Malaysia		
	72	Rapier launchers
Mali		
	8	SA-9 launchers
	16	SA-3 launchers
Mauritania		
	4	SA-9 launchers
Morocco		
	37	Chaparral launchers
Mozambique		
	32	SA-9 launchers
	12	SA-3 launchers
	24	SA-6 launchers
	32	SA-8 launchers
	12	SA-2 launchers
Netherlands		
	72	Hawk launchers
	20	Patriot launchers
	27	Nike-Hercules launchers
Nicaragua		
	12	SA-9 launchers

Country	Quantity (1991)	Surface-to-Air Missile
Nigeria		
	16	Roland launchers
North Korea		
	32	SA-3 launchers
	270	SA-2 launchers
	24	SA-5 launchers
Norway		
	54	I-Hawk launchers
	36	Nike-Hercules launchers
Oman		
	28	Rapier launchers
Pakistan		
	24	Crotale (Firing Unit)
	6	HQ-2 launchers (SA-2)
Peru		
	12	SA-3 launchers
Poland		
	200	SA-9 launchers
	20	SA-11 launchers
	60	SA-13 launchers
	200	SA-3 launchers
	120	SA-6 launchers
	60	SA-8 launchers
	240	SA-2 launchers
	24	SA-4 launchers
Portugal		
	5	Chaparral launchers
	1	Hawk launchers
Qatar		
	9	Roland launchers
	18	Rapier launchers
	5	Tigercat launchers
Romania		
	40	SA-9 launchers
	60	SA-6 launchers
	108	SA-2 launchers
Saudi Arabia		
	16	Shahine I launchers
	220	Shahine II launchers
	126	Hawk launchers
	90	Improved Hawk launchers
		Patriot launchers
Singapore		
	12	Rapier launchers
	18	Hawk launchers
	28	Bloodhound launcher
Somalia		
	8	SA-3 launchers
	12	SA-6 launchers
	42	SA-2 launchers
South Africa		
	16	SA-9 launchers
	14	Crotale (firing units)

Country	Quantity (1991)	Surface-to-Air Missile
South Korea		
	168	Hawk launchers
	90	Nike Hercules launchers
Spain		
	18	Roland launchers
	13	Skyguard/Aspides launchers
	48	Hawk launchers
	9	Nike Hercules launchers
Sudan		
	30	SA-2 launchers
Sweden		
	12	Hawk launchers
Switzerland		
	60	Rapier launchers
	64	Bloodhound launchers
Syria		
	150	SA-9 launcher
	20	SA-11 launcher
	60	SA-13 launcher
	160	SA-3 launcher
	60	SA-8 launcher
	216	SA-6 launcher
	138	SA-2 launcher
	24	SA-5 launcher
Taiwan		
	92	Chaparral launchers
	78	Hawk launchers
	96	Nike Hercules launchers
Tanzania		
	40	SA-9 launchers
	12	SA-3 launchers
	12	SA-6 launchers
Thailand		
	20	Crotale NG launcher
	4	Aspide/Spada launchers
	78	Hawk launchers
	36	Nike Hercules launcher
Tunisia		
	26	Chaparral launchers
Turkey		
	72	Rapier launchers
	33	Patriot launchers
	72	Nike Hercules launcher
United Arab Emirates		
	9	Crotale launchers
	42	Hawk launchers
United Kingdom		
	72	Tracked Rapier
	32	Bloodhound 2 launchers

Country	Quantity (1991)	Surface-to-Air Missile
United States		
	27	Roland launchers
	74	Rapier launchers
	4	ADATS launchers
	150	Chaparral launchers
	4	Aspide/Spada Battery
	133	Hawk launchers
	432	Patriot launchers
Vietnam		
	450	SA-9 launchers
	160	SA-3 launchers
	80	SA-6 launchers
	360	SA-2 launchers
Venezuela		
	8	Roland launchers
Yemen		
	20	SA-9 launcher
	12	SA-3 launcher
	40˙	SA-6 launcher
	66	SA-2 launcher
Yugoslavia		
	24	SA-3 launchers
	80	SA-6 launchers
	48	SA-2 launchers
Zambia		
	12	Rapier launchers
	3	Tigercat launchers

3. Aircraft Data

	Afghanistan	Albania	Algeria	Angola	Argentina	Australia
FIGHTERS						
Front-line fighters						
F-14	-	-	-	-	-	-
F-15	-	-	-	-	-	-
F-16	-	-	-	-	-	-
F-18	-	-	-	-	-	71
F-117						
Harrier	-	-	-	-	-	-
MiG-29	-	-	-	-	-	-
MiG-31						
Mirage 2000	-	-	-	-	-	-
Su-24	-	-	6	-	-	-
Su-27						
Tornado	-	-	-	-	-	-
Capable fighters						
A-10A	-	-	-	-	-	-
A-37 Viggen	-	-	-	-	-	-
A-6	-	-	-	-	-	-
A-7	-	-	-	-	-	-
F-4	-	-	-	-	-	-
F-111	-	-	-	-	-	22
J-8						
Jaguar	-	-	-	-	-	-
MiG-23/27	30	-	62	51	-	-
MiG-25	-	-	20	-	-	-
Mirage F.1	-	-	-	-	-	-
Su-25	-	-	-	10	-	-
Super Entendard IV	-	-	-	-	11	-
Yak-38						
Old fighters						
A-4	-	-	-	-	16	-
Buccaneer	-	-	-	-	-	-
Crusader	-	-	-	-	-	-
F-5	-	-	-	-	-	-
F-104	-	-	-	-	-	-
F-35 Draken	-	-	-	-	-	-
F-86F						
G-91	-	-	-	-	-	-
Hunter	-	-	-	-	-	-
Kfir	-	-	-	-	-	-
MiG-15 (J-2)	-	20	3	3	-	-
MiG-17 (J-5)	-	8	55	25	-	-
MiG-19 (J-6)	-	30	-	-	-	-
MiG-21 (J-7)	118	20	98	41	-	-
Mirage III	-	-	-	-	58	-
Mirage 5	-	-	-	-	8	-
Ouragon						
Q-5	-	-	-	-	-	-
Su-7	80	-	-	-	-	-
Su-15						

	Afghanistan	Albania	Algeria	Angola	Argentina	Australia
Old fighters						
Su-17/20/22	-	-	-	15	-	-
Super Mystere B2						
Fighter summary						
Front-line	-	-	6	-	-	71
Less capable	30	-	82	61	11	22
Old	198	78	156	84	82	-
Total fighters	228	78	244	145	93	93
BOMBERS	-	-	-	-	6	-
SUPPORT AIRCRAFT						
Airlift	42	19	44	68	97	54
Armed trainers	25	-	24	-	21	46
Utility trainers	25	6	73	9	95	95
Counterinsurgency	-	-	-	16	78	4
MR/ASW	-	-	2	3	9	19
Tankers	-	-	-	-	4	4
Other	1	-	-	-	20	8
Total support aircraft	93	25	143	96	304	230

	Austria	Bahamas	Bahrain	Bangladesh	Belgium	Belize
FIGHTERS						
Front-line fighters						
F-14	-	-	-	-	-	-
F-15	-	-	-	-	-	-
F-16	-	-	12	-	120	-
F-18	-	-	-	-	-	-
F-117						
Harrier	-	-	-	-	-	-
MiG-29	-	-	-	-	-	-
MiG-31						
Mirage 2000	-	-	-	-	-	-
Su-24	-	-	-	-	-	-
Su-27						
Tornado	-	-	-	-	-	-
Capable fighters						
A-10A	-	-	-	-	-	-
A-37 Viggen	-	-	-	-	-	-
A-6	-	-	-	-	-	-
A-7	-	-	-	-	-	-
F-4	-	-	-	-	-	-
F-111	-	-	-	-	-	-
J-8						
Jaguar	-	-	-	-	-	-
MiG-23/27	-	-	-	-	-	-
MiG-25	-	-	-	-	-	-
Mirage F.1	-	-	-	-	-	-
Su-25	-	-	-	-	-	-
Super Entendard IV	-	-	-	-	-	-
Yak-38						
Old fighters						
A-4	-	-	-	-	-	-
Buccaneer	-	-	-	-	-	-
Crusader	-	-	-	-	-	-
F-5	-	-	12	-	-	-
F-104	-	-	-	-	-	-
F-35 Draken	24	-	-	-	-	-
F-86F						
G-91	-	-	-	-	-	-
Hunter	-	-	-	-	-	-
Kfir	-	-	-	-	-	-
MiG-15 (J-2)	-	-	-	4	-	-
MiG-17 (J-5)	-	-	-	-	-	-
MiG-19 (J-6)	-	-	-	38	-	-
MiG-21 (J-7)	-	-	-	39	-	-
Mirage III	-	-	-	-	-	-
Mirage 5	-	-	-	-	65	-
Ouragon						
Q-5	-	-	-	16	-	-
Su-7	-	-	-	15	-	-
Su-15						
Su-17/20/22	-	-	-	-	-	-
Super Mystere B2						

	Austria	Bahamas	Bahrain	Bangladesh	Belgium	Belize
Fighter summary						
Front-line	-	-	12	-	120	-
Less capable	-	-	-	-	-	-
Old	24	-	12	112	65	-
Total fighters	24	-	24	112	185	-
BOMBERS	-	-	-	-	-	-
SUPPORT AIRCRAFT						
Airlift	13	5	2	6	52	-
Armed trainers	38	-	-	-	31	-
Utility trainers	38	-	-	8	18	-
Counterinsurgency	15	-	-	-	-	-
MR/ASW	-	-	-	-	-	3
Tankers	-	-	-	-	-	-
Other	-	-	-	-	-	-
Total support aircraft	104	5	2	14	101	3

	Benin	Bolivia	Bophuthat-swana	Botswana	Brazil	Brunei
FIGHTERS						
Front-line fighters						
F-14	-	-	-	-	-	-
F-15	-	-		-	-	-
F-16	-	-	-	-	-	-
F-18	-	-	-	-	-	-
F-117						
Harrier	-	-	-	-	-	-
MiG-29	-	-	-	-	-	-
MiG-31						
Mirage 2000	-	-	-	-	-	-
Su-24	-	-	-	-	-	-
Su-27						
Tornado	-	-	-	-	-	-
Capable fighters						
A-10A	-	-	-	-	-	
A-37 Viggen	-	-	-	-	-	-
A-6	-	-	-	-	-	-
A-7	-	-	-	-	-	-
F-4	-	-	-	-	-	-
F-111	-	-	-	-	-	-
J-8						
Jaguar	-	-	-	-	-	-
MiG-23/27	-	-	-	-	-	-
MiG-25	-	-	-	-	-	-
Mirage F.1	-	-	-	-	-	-
Su-25	-	-	-	-	-	-
Super Entendard IV	-	-	-	-	-	-
Yak-38						
Old fighters						
A-4	-	-	-	-	-	-
Buccaneer	-	-	-	-	-	-
Crusader	-	-	-	-	-	-
F-5	-	-	-	-	57	-
F-104	-	-	-	-	-	-
F-35 Draken	-	-	-	-	-	-
F-86F		4				
G-91	-	-	-	-	-	-
Hunter	-	-	-	-	-	-
Kfir	-	-	-	-	-	-
MiG-15 (J-2)	-	-	-	-	-	-
MiG-17 (J-5)	-	-	-	-	-	-
MiG-19 (J-6)	-	-	-	-	-	-
MiG-21 (J-7)	-	-	-	-	-	-
Mirage III	-	-	-	-	18	-
Mirage 5	-	-	-	-	-	-
Ouragon						
Q-5	-	-	-	-	-	-
Su-7	-	-	-	-	-	-
Su-15						
Su-17/20/22	-	-	-	-	-	-
Super Mystere B2						

	Benin	Bolivia	Bophuthat-swana	Botswana	Brazil	Brunei
Fighter summary						
Front-line	-	-	-	-	-	-
Less capable	-	-	-	-	-	-
Old	-	4	-	-	75	-
Total fighters	-	4	-	-	75	-
BOMBERS	-	-	-	-	-	-
SUPPORT AIRCRAFT						
Airlift	10	51	4	15	257	-
Armed trainers	-	21	-	8	21	-
Utility trainers	-	47	1	-	279	4
Counterinsurgency	-	12	-	5	128	-
MR/ASW	-	-	-	-	11	-
Tankers	-	-	-	-	6	-
Other	-	6	-	-	19	-
Total support aircraft	10	137	5	28	721	4

	Bulgaria	Burkina Faso	Burundi	Cambodia	Cameroon	Canada
FIGHTERS						
Front-line fighters						
F-14	-	-	-	-	-	-
F-15	-	-	-	-	-	-
F-16	-	-	-	-	-	-
F-18	-	-	-	-	-	127
F-117						
Harrier	-	-	-	-	-	-
MiG-29	22	-	-	-	-	-
MiG-31						
Mirage 2000	-	-	-	-	-	-
Su-24	-	-	-	-	-	-
Su-27						
Tornado	-	-	-	-	-	-
Capable fighters						
A-10A	-	-	-	-	-	-
A-37 Viggen	-	-	-	-	-	-
A-6	-	-	-	-	-	-
A-7	-	-	-	-	-	-
F-4	-	-	-	-	-	-
F-111	-	-	-	-	-	-
J-8						
Jaguar	-	-	-	-	-	-
MiG-23/27	42	-	-	-	-	-
MiG-25	5	-	-	-	-	-
Mirage F.1	-	-	-	-	-	-
Su-25	39	-	-	-	-	-
Super Entendard IV	-	-	-	-	-	-
Yak-38						
Old fighters						
A-4	-	-	-	-	-	-
Buccaneer	-	-	-	-	-	-
Crusader	-	-	-	-	-	-
F-5	-	-	-	-	-	118
F-104	-	-	-	-	-	-
F-35 Draken	-	-	-	-	-	-
F-86F						
G-91	-	-	-	-	-	-
Hunter	-	-	-	-	-	-
Kfir	-	-	-	-	-	-
MiG-15 (J-2)	-	-	-	-	-	-
MiG-17 (J-5)	-	-	-	-	-	-
MiG-19 (J-6)	-	-	-	-	-	-
MiG-21 (J-7)	138	8	-	17	-	-
Mirage III	-	-	-	-	-	-
Mirage 5	-	-	-	-	-	-
Ouragon						
Q-5	-	-	-	-	-	-
Su-7	-	-	-	-	-	-
Su-15						
Su-17/20/22	20	-	-	-	-	-
Super Mystere B2						

	Bulgaria	Burkina Faso	Burundi	Cambodia	Cameroon	Canada
Fighter summary						
Front-line	22	-	-	-	-	127
Less capable	86	-	-	-	-	-
Old	158	8	-	17	-	118
Total fighters	266	8	-	17	-	245
BOMBERS	-	-	-	-	-	-
SUPPORT AIRCRAFT						
Airlift	16	14	5	3	8	74
Armed trainers	107	-	-	-	16	-
Utility trainers	123	-	7	-	-	170
Counterinsurgency	-	10	3	-	-	-
MR/ASW	-	-	-	-	2	40
Tankers	-	-	-	-	-	4
Other	2	-	-	-	-	8
Total support aircraft	248	14	15	3	26	256

	Cape Verde	Cent. African Republic	Chad	Chile	China	Ciskei (S. Africa)
FIGHTERS						
Front-line fighters						
F-14	-	-	-	-	-	-
F-15	-	-	-	-	-	-
F-16	-	-	-	-	-	-
F-18	-	-	-	-	-	-
F-117						
Harrier	-	-	-	-	-	-
MiG-29	-	-	-	-	-	-
MiG-31						
Mirage 2000	-	-	-	-	-	-
Su-24	-	-	-	-	-	-
Su-27					24	
Tornado	-	-	-	-	-	-
Capable fighters						
A-10A	-	-	-	-	-	-
A-37 Viggen	-	-	-	-	-	-
A-6	-	-	-	-	-	-
A-7	-	-	-	-	-	-
F-4	-	-	-	-	-	-
F-111	-	-	-	-	-	-
J-8					150	
Jaguar	-	-	-	-	-	-
MiG-23/27	-	-	-	-	-	-
MiG-25	-	-	-	-	-	-
Mirage F.1	-	-	-	-	-	-
Su-25	-	-	-	-	-	-
Super Entendard IV	-	-	-	-	-	-
Yak-38						
Old fighters						
A-4	-	-	-	-	-	-
Buccaneer	-	-	-	-	-	-
Crusader	-	-	-	-	-	-
F-5	-	-	-	16	-	-
F-104	-	-	-	-	-	-
F-35 Draken	-	-	-	-	-	-
F-86F						
G-91	-	-	-	-	-	-
Hunter	-	-	-	33	-	-
Kfir	-	-	-	-		-
MiG-15 (J-2)	-	-	-	-	100	-
MiG-17 (J-5)	-	-	-	-	1790	-
MiG-19 (J-6)	-	-	-	-	3300	-
MiG-21 (J-7)	-	-	-	-	500	-
Mirage III	-	-	-	-	-	-
Mirage 5	-	-	-	15	-	-
Ouragon						
Q-5	-	-	-	-	750	-
Su-7	-	-	-	-	-	-
Su-15						
Su-17/20/22	-	-	-	-	-	-
Super Mystere B2						

	Cape Verde	Cent. African Republic	Chad	Chile	China	Ciskei (S. Africa)
Fighter summary						
Front-line	-	-	-	-	24	-
Less capable	-	-	-	-	150	-
Old	-	-	-	64	6440	-
Total fighters	-	-	-	64	6614	-
BOMBERS	-	-	-	-	570	-
SUPPORT AIRCRAFT						
Airlift	2	19	17	65	730	5
Armed trainers	-	-	-	51	-	-
Utility trainers	-	-	-	156	1000	-
Counterinsurgency	-	-	4	20	-	-
MR/ASW	-	-	-	8	15	-
Tankers	-	-	-	-	-	-
Other	-	-	-	3	40	-
Total support aircraft	2	19	21	283	1745	5

	Colombia	Congo	Costa Rica	Cote d'Ivoire	Cuba	Cyprus
FIGHTERS						
Front-line fighters						
F-14	-	-	-	-	-	-
F-15	-	-	-	-	-	-
F-16	-	-	-	-	-	-
F-18	-	-	-	-	-	-
F-117						
Harrier	-	-	-	-	-	-
MiG-29	-	-	-	-	8	-
MiG-31						
Mirage 2000	-	-	-	-	-	-
Su-24	-	-	-	-	-	-
Su-27						
Tornado	-	-	-	-	-	-
Capable fighters						
A-10A	-	-	-	-	-	-
A-37 Viggen	-	-	-	-	-	-
A-6	-	-	-	-	-	-
A-7	-	-	-	-	-	-
F-4	-	-	-	-	-	-
F-111	-	-	-	-	-	-
J-8						
Jaguar	-	-	-	-	-	-
MiG-23/27	-	-	-	-	64	-
MiG-25	-	-	-	-	-	-
Mirage F.1	-	-	-	-	-	-
Su-25	-	-	-	-	-	-
Super Entendard IV	-	-	-	-	-	-
Yak-38						
Old fighters						
A-4	-	-	-	-	-	-
Buccaneer	-	-	-	-	-	-
Crusader	-	-	-	-	-	-
F-5	-	-	-	-	-	-
F-104	-	-	-	-	-	-
F-35 Draken	-	-	-	-	-	-
F-86F						
G-91	-	-	-	-	-	-
Hunter	-	-	-	-	-	-
Kfir	13	-	-	-	-	-
MiG-15 (J-2)	-	1	-	-	30	-
MiG-17 (J-5)	-	20	-	-	-	-
MiG-19 (J-6)	-	-	-	-	-	-
MiG-21 (J-7)	-	12	-	-	90	-
Mirage III	-	-	-	-	-	-
Mirage 5	15	-	-	-	-	-
Ouragon						
Q-5	-	-	-	-	-	-
Su-7	-	-	-	-	-	-
Su-15						
Su-17/20/22	-	-	-	-	-	-
Super Mystere B2						

	Colombia	Congo	Costa Rica	Cote d'Ivoire	Cuba	Cyprus
Fighter summary						
Front-line	-	-	-	-	8	-
Less capable	-	-	-	-	64	-
Old	28	33	-	-	120	-
Total fighters	28	33	-	-	192	-
BOMBERS	-	-	-	-	-	-
SUPPORT AIRCRAFT						
Airlift	67	8	12	6	69	4
Armed trainers	42	4	-	6	25	-
Utility trainers	50	4	-	8	45	-
Counterinsurgency	4	-	3	-	-	-
MR/ASW	-	-	-	-	-	-
Tankers	-	-	-	-	-	-
Other	-	-	-	-	-	-
Total support aircraft	163	16	15	20	139	4

	Czecho-slovakia	Denmark	Djibouti	Dominican Republic	Ecuador	Egypt
FIGHTERS						
Front-line fighters						
F-14	-	-	-	-	-	-
F-15	-	-	‎-	-	-	-
F-16	-	63	-	-	-	79
F-18	-	-	-	-	-	-
F-117						
Harrier	-	-	-	-	-	-
MiG-29	20	-	-	-	-	-
MiG-31						
Mirage 2000	-	-	-	-	-	19
Su-24	-	-	-	-	-	-
Su-27						
Tornado	-	-	-	-	-	-
Capable fighters						
A-10A	-	-	-	-	-	-
A-37 Viggen	-	-	-	-	-	-
A-6	-	-	-	-	-	-
A-7	-	-	-	-	-	-
F-4	-	-	-	-	-	33
F-111	-	-	-	-	-	-
J-8						
Jaguar	-	-	-	-	9	-
MiG-23/27	65	-	-	-	-	-
MiG-25	-	-	-	-	-	-
Mirage F.1	-	-	-	-	14	-
Su-25	38	-	-	-	-	-
Super Entendard IV	-	-	-	-	-	-
Yak-38						
Old fighters						
A-4	-	-	-	-	-	-
Buccaneer	-	-	-	-	-	-
Crusader	-	-	-	-	-	-
F-5	-	-	-	-	-	-
F-104	-	-	-	-	-	-
F-35 Draken	-	43	-	-	-	-
F-86F						
G-91	-	-	-	-	-	-
Hunter	-	-	-	-	-	-
Kfir	-	-	-	-	10	-
MiG-15 (J-2)	-	-	-	-	-	-
MiG-17 (J-5)	-	-	-	-	-	-
MiG-19 (J-6)	-	-	-	-	-	92
MiG-21 (J-7)	116	-	-	-	-	149
Mirage III	-	-	-	-	-	-
Mirage 5	-	-	-	-	-	81
Ouragon						
Q-5	-	-	-	-	-	-
Su-7	-	-	-	-	-	-
Su-15						
Su-17/20/22	58	-	-	-	-	-
Super Mystere B2						

	Czecho-slovakia	Denmark	Djibouti	Dominican Republic	Ecuador	Egypt
Fighter summary						
Front-line	20	63	-	-	-	98
Less capable	103	-	-	-	23	33
Old	174	43	-	-	10	322
Total fighters	297	106	-	-	33	453
BOMBERS	-	-	-	-	-	-
SUPPORT AIRCRAFT						
Airlift	32	20	6	11	37	20
Armed trainers	100	-	-	8	37	92
Utility trainers	150	23	-	9	56	154
Counterinsurgency	-	-	-	5	9	-
MR/ASW	-	-	-	-	-	-
Tankers	-	-	-	-	-	-
Other	6	-	-	-	2	6
Total support aircraft	288	43	6	33	141	272

	El Salvador	Equatorial Guinea	Ethiopia	Finland	France	Gabon
FIGHTERS						
Front-line fighters						
F-14	-	-	-	-	-	-
F-15	-	-	-	-	-	-
F-16	-	-	-	-	-	-
F-18	-	-	-	-	-	-
F-117						
Harrier	-	-	-	-	-	-
MiG-29	-	-	-	-	-	-
MiG-31						
Mirage 2000	-	-	-	-	151	-
Su-24	-	-	-	-	-	-
Su-27						
Tornado	-	-	-	-	-	-
Capable fighters						
A-10A	-	-	-	-	-	-
A-37 Viggen	-	-	-	-	-	-
A-6	-	-	-	-	-	-
A-7	-	-	-	-	-	-
F-4	-	-	-	-	-	-
F-111	-	-	-	-	-	-
J-8						
Jaguar	-	-	-	-	156	-
MiG-23/27	-	-	18	-	-	-
MiG-25	-	-	-	-	-	-
Mirage F.1	-	-	-	-	207	-
Su-25	-	-	-	-	-	-
Super Entendard IV	-	-	-	-	59	-
Yak-38						
Old fighters						
A-4	-	-	-	-	-	-
Buccaneer	-	-	-	-	-	-
Crusader	-	-	-	-	19	-
F-5	-	-	-	-	-	-
F-104	-	-	-	-	-	-
F-35 Draken	-	-	-	40	-	-
F-86F						
G-91	-	-	-	-	-	-
Hunter	-	-	-	-	-	-
Kfir	-	-	-	-	-	-
MiG-15 (J-2)	-	-	-	-	-	-
MiG-17 (J-5)	-	-	-	-	-	-
MiG-19 (J-6)	-	-	-	-	-	-
MiG-21 (J-7)	-	-	50	24	-	-
Mirage III	-	-	-	-	113	-
Mirage 5	-	-	-	-	39	9
Ouragon	8					
Q-5	-	-	-	-	-	-
Su-7	-	-	-	-	-	-
Su-15						
Su-17/20/22	-	-	-	-	-	-
Super Mystere B2						

	El Salvador	Equatorial Guinea	Ethiopia	Finland	France	Gabon
Fighter summary						
Front-line	-	-	-	-	151	-
Less capable	-	-	18	-	422	-
Old	8	-	50	64	171	9
Total fighters	8	-	68	64	744	9
BOMBERS	-	-	-	-	60	-
SUPPORT AIRCRAFT						
Airlift	33	1	14	18	192	13
Armed trainers	11	-	14	50	159	6
Utility trainers	1	-	14	28	317	-
Counterinsurgency	17	-	-	-	-	4
MR/ASW	-	-	-	-	63	1
Tankers	-	-	-	-	11	-
Other	-	-	-	3	2	-
Total support aircraft	62	1	42	99	744	24

FIGHTERS	Germany	Ghana	Greece	Guatemala	Guinea	Guinea-Bissau
Front-line fighters						
F-14	-	-	-	-	-	-
F-15	-	-	-	-	-	-
F-16	-	-	40	-	-	-
F-18	-	-	-	-	-	-
F-117						
Harrier	-	-	-	-	-	-
MiG-29	24	-	-	-	-	-
MiG-31						
Mirage 2000	-	-	32	-	-	-
Su-24	-	-	-	-	-	-
Su-27						
Tornado	313	-	-	-	-	-
Capable fighters						
A-10A	-	-	-	-	-	-
A-37 Viggen	-	-	-	-	-	-
A-6	-	-	-	-	-	-
A-7	-	-	61	-	-	-
F-4	231	-	59	-	-	-
F-111	-	-	-	-	-	-
J-8						
Jaguar	-	-	-	-	-	-
MiG-23/27	58	-	-	-	-	-
MiG-25	-	-	-	-	-	-
Mirage F.1	-	-	40	-	-	-
Su-25	-	-	-	-	-	-
Super Entendard IV	-	-	-	-	-	-
Yak-38						
Old fighters						
A-4	-	-	-	-	-	-
Buccaneer	-	-	-	-	-	-
Crusader	-	-	-	-	-	-
F-5	-	-	106	-	-	-
F-104	-	-	120	-	-	-
F-35 Draken	-	-	-	-	-	-
F-86F						
G-91	-	-	-	-	-	-
Hunter	-	-	-	-	-	-
Kfir	-	-	-	-	-	-
MiG-15 (J-2)	-	-	-	-	2	-
MiG-17 (J-5)	-	-	-	-	4	3
MiG-19 (J-6)	-	-	-	-	-	-
MiG-21 (J-7)	251	-	-	-	8	-
Mirage III	-	-	-	-	-	-
Mirage 5	-	-	-	-	-	-
Ouragon						
Q-5	-	-	-	-	-	-
Su-7	-	-	-	-	-	-
Su-15						
Su-17/20/22	27	-	-	-	-	-
Super Mystere B2						

	Germany	Ghana	Greece	Guatemala	Guinea	Guinea-Bissau
Fighter summary						
Front-line	337	-	72	-	-	-
Less capable	289	-	160	-	-	-
Old	278	-	226	-	14	3
Total fighters	904	-	458	-	14	3
BOMBERS	-	-	-	-	-	-
SUPPORT AIRCRAFT						
Airlift	239	13	66	28	6	-
Armed trainers	164	18	81	8	3	-
Utility trainers	76	22	143	6	9	-
Counterinsurgency	-	-	-	8	-	-
MR/ASW	14	-	12	-	-	-
Tankers	-	-	-	-	-	-
Other	25	-	-	-	-	-
Total support aircraft	518	53	302	50	18	-

FIGHTERS	Guyana	Haiti	Honduras	Hungary	India	Indonesia
Front-line fighters						
F-14	-	-	-	-	-	-
F-15	-	-	-	-	-	-
F-16	-	-	-	-	-	12
F-18	-	-	-	-	-	-
F-117						
Harrier	-	-	-	-	-	-
MiG-29	-	-	-	-	59	-
MiG-31						
Mirage 2000	-	-	-	-	36	-
Su-24	-	-	-	-	-	-
Su-27						
Tornado	-	-	-	-	-	-
Capable fighters						
A-10A	-	-	-	-	-	-
A-37 Viggen	-	-	-	-	-	-
A-6	-	-	-	-	-	-
A-7	-	-	-	-	-	-
F-4	-	-	-	-	-	-
F-111	-	-	-	-	-	-
J-8						
Jaguar	-	-	-	-	93	-
MiG-23/27	-	-	-	12	136	-
MiG-25	-	-	-	-	8	-
Mirage F.1	-	-	-	-	-	-
Su-25	-	-	-	-	-	-
Super Entendard IV	-	-	-	-	-	-
Yak-38						
Old fighters						
A-4	-	-	-	-	-	28
Buccaneer	-	-	-	-	-	-
Crusader	-	-	-	-	-	-
F-5	-	-	12	-	-	14
F-104	-	-	-	-	-	-
F-35 Draken	-	-	-	-	-	-
F-86F						
G-91	-	-	-	-	-	-
Hunter	-	-	-	-	20	-
Kfir	-	-	-	-	-	-
MiG-15 (J-2)	-	-	-	-	-	-
MiG-17 (J-5)	-	-	-	-	-	-
MiG-19 (J-6)	-	-	-	-	-	-
MiG-21 (J-7)	-	-	-	119	294	-
Mirage III	-	-	-	-	-	-
Mirage 5	-	-	-	-	-	-
Ouragon						
Q-5	-	-	-	-	-	-
Su-7	-	-	-	-	-	-
Su-15						
Su-17/20/22	-	-	-	11	-	-
Super Mystere B2			8			

	Guyana	Haiti	Honduras	Hungary	India	Indonesia
Fighter summary						
Front-line	-	-	-	-	95	12
Less capable	-	-	-	12	237	-
Old	-	-	20	130	314	42
Total fighters	-	-	20	142	646	54
BOMBERS	-	-	-	-	-	-
SUPPORT AIRCRAFT						
Airlift	8	2	36	16	243	61
Armed trainers	-	-	15	-	-	15
Utility trainers	-	10	26	-	346	90
Counterinsurgency	-	-	-	-	-	12
MR/ASW	-	-	-	-	26	27
Tankers	-	-	-	-	-	2
Other	-	-	-	-	3	-
Total support aircraft	8	2	77	16	618	117

FIGHTERS	Iran	Iraq	Ireland	Israel	Italy	Jamaica
Front-line fighters						
F-14	60	-	-	-	-	-
F-15	-	-	-	47	-	-
F-16	-	-	-	149	-	-
F-18	-	-	-	-	-	-
F-117						
Harrier	-	-	-	-	-	-
MiG-29	20	-	-	-	-	-
MiG-31						
Mirage 2000	-	-	-	-	-	-
Su-24	-	-	-	-	-	-
Su-27						
Tornado	-	-	-	-	96	-
Capable fighters						
A-10A	-	-	-	-	-	-
A-37 Viggen	-	-	-	-	-	-
A-6	-	-	-	-	-	-
A-7	-	-	-	-	-	-
F-4	60	-	-	125	-	-
F-111	-	-	-	-	-	-
J-8						
Jaguar	-	-	-	-	-	-
MiG-23/27	-	130	-	-	-	-
MiG-25	-	4	-	-	-	-
Mirage F.1	-	125	-	-	-	-
Su-25	-	-	-	-	-	-
Super Entendard IV	-	-	-	-	-	-
Yak-38						
Old fighters						
A-4	-	-	-	162	-	-
Buccaneer	-	-	-	-	-	-
Crusader	-	-	-	-	-	-
F-5	60	-	-	-	-	-
F-104	-	-	-	-	200	-
F-35 Draken	-	-	-	-	-	-
F-86F						
G-91	-	-	-	-	129	-
Hunter	-	-	-	-	-	-
Kfir	-	-	-	175	-	-
MiG-15 (J-2)	-	-	-	-	-	-
MiG-17 (J-5)	-	-	-	-	-	-
MiG-19 (J-6)	-	-	-	-	-	-
MiG-21 (J-7)	-	-	-	-	-	-
Mirage III	-	-	-	-	-	-
Mirage 5	-	-	-	-	-	-
Ouragon						
Q-5	-	-	-	-	-	-
Su-7	-	-	-	-	-	-
Su-15						
Su-17/20/22	-	130	-	-	-	-
Super Mystere B2						

	Iran	Iraq	Ireland	Israel	Italy	Jamaica
Fighter summary						
Front-line	80	-	-	196	96	-
Less capable	60	259	-	125	-	-
Old	60	130	-	337	329	-
Total fighters	200	389	-	658	425	-
BOMBERS	-	6	-	-	-	-
SUPPORT AIRCRAFT						
Airlift	55	-	10	145	111	5
Armed trainers	7	50	6	-	197	-
Utility trainers	118	50	-	121	39	-
Counterinsurgency	-	-	8	-	-	-
MR/ASW	5	-	2	5	18	-
Tankers	4	-	-	7	1	-
Other	-	-	1	30	28	-
Total support aircraft	189	50	27	278	394	5

FIGHTERS	Japan	Jordan	Kenya	Kuwait	Laos	Lebanon
Front-line fighters						
F-14	-	-	-	-	-	-
F-15	147	-	-	-	-	-
F-16	-	-	-	-	-	-
F-18	-	-	-	-	-	-
F-117						
Harrier	-	-	-	-	-	-
MiG-29	-	-	-	-	-	-
MiG-31						
Mirage 2000	-	-	-	-	-	-
Su-24	-	-	-	-	-	-
Su-27						
Tornado	-	-	-	-	-	-
Capable fighters						
A-10A	-	-	-	-	-	-
A-37 Viggen	-	-	-	-	-	-
A-6	-	-	-	-	-	-
A-7	-	-	-	-	-	-
F-4	158	-	-	-	-	-
F-111	-	-	-	-	-	-
J-8						
Jaguar	-	-	-	-	-	-
MiG-23/27	-	-	-	-	-	-
MiG-25	-	-	-	-	-	-
Mirage F.1	-	32	-	15	-	-
Su-25	-	-	-	-	-	-
Super Entendard IV	-	-	-	-	-	-
Yak-38						
Old fighters						
A-4	-	-	-	19	-	-
Buccaneer	-	-	-	-	-	-
Crusader	-	-	-	-	-	-
F-5	-	81	11	-	-	-
F-104	-	-	-	-	-	-
F-35 Draken	-	-	-	-	-	-
F-86F						
G-91	-	-	-	-	-	-
Hunter	-	-	-	-	-	3
Kfir	-	-	-	-	-	-
MiG-15 (J-2)	-	-	-	-	-	-
MiG-17 (J-5)	-	-	-	-	-	-
MiG-19 (J-6)	-	-	-	-	-	-
MiG-21 (J-7)	-	-	-	-	34	-
Mirage III	-	-	-	-	-	-
Mirage 5	-	-	-	-	-	-
Ouragon						
Q-5	-	-	-	-	-	-
Su-7	-	-	-	-	-	-
Su-15						
Su-17/20/22	-	-	-	-	-	-
Super Mystere B2						

	Japan	Jordan	Kenya	Kuwait	Laos	Lebanon
Fighter summary						
Front-line	147	-	-	-	-	-
Less capable	158	32	-	15	-	-
Old	-	81	11	19	34	3
Total fighters	305	113	11	34	34	3
BOMBERS	-	-	-	-	-	-
SUPPORT AIRCRAFT						
Airlift	110	13	8	-	9	-
Armed trainers	208	-	17	-	-	-
Utility trainers	443	49	8	-	-	6
Counterinsurgency	-	-	-	-	-	-
MR/ASW	129	-	-	-	-	-
Tankers	-	-	-	-	-	-
Other	31	-	-	-	-	-
Total support aircraft	921	62	33	-	9	6

	Libya	Madagascar	Malawi	Malaysia	Mali	Mauritania
FIGHTERS						
Front-line fighters						
F-14	-	-	-	-	-	-
F-15	-	-	-	-	-	-
F-16	-	-	-	-	-	-
F-18	-	-	-	-	-	-
F-117						
Harrier	-	-	-	-	-	-
MiG-29	-	-	-	-	-	-
MiG-31						
Mirage 2000	-	-	-	-	-	-
Su-24	10	-	-	-	-	-
Su-27						
Tornado	-	-	-	-	-	-
Capable fighters						
A-10A	-	-	-	-	-	-
A-37 Viggen	-	-	-	-	-	-
A-6	-	-	-	-	-	-
A-7	-	-	-	-	-	-
F-4	-	-	-	-	-	-
F-111	-	-	-	-	-	-
J-8						
Jaguar	-	-	-	-	-	-
MiG-23/27	140	-	-	-	-	-
MiG-25	65	-	-	-	-	-
Mirage F.1	26	-	-	-	-	-
Su-25	-	-	-	-	-	-
Super Entendard IV	-	-	-	-	-	-
Yak-38						
Old fighters						
A-4	-	-	-	32	-	-
Buccaneer	-	-	-	-	-	-
Crusader	-	-	-	-	-	-
F-5	-	-	-	21	-	-
F-104	-	-	-	-	-	-
F-35 Draken	-	-	-	-	-	-
F-86F						
G-91	-	-	-	-	-	-
Hunter	-	-	-	-	-	-
Kfir	-	-	-	-	-	-
MiG-15 (J-2)	-	-	-	-	1	-
MiG-17 (J-5)	-	4	-	-	5	-
MiG-19 (J-6)	-	-	-	-	-	-
MiG-21 (J-7)	50	8	-	-	11	-
Mirage III	-	-	-	-	-	-
Mirage 5	38	-	-	-	-	-
Ouragon	-					
Q-5	-	-	-	-	-	-
Su-7	-	-	-	-	-	-
Su-15						
Su-17/20/22	45	-	-	-	-	-
Super Mystere B2						

	Libya	Madagascar	Malawi	Malaysia	Mali	Mauritania
Fighter summary						
Front-line	10	-	-	-	-	-
Less capable	231	-	-	-	-	-
Old	133	12	-	53	17	-
Total fighters	374	12	-	53	17	-
BOMBERS	9	-	-	-	-	-
SUPPORT AIRCRAFT						
Airlift	74	15	8	46	6	4
Armed trainers	119	-	-	10	6	-
Utility trainers	238	4	-	52	12	-
Counterinsurgency	30	-	-	-	-	7
MR/ASW	-	-	-	3	-	2
Tankers	-	-	-	-	-	-
Other	-	-	-	-	-	-
Total support aircraft	431	19	8	111	24	13

62

FIGHTERS	Mexico	Mongolia	Morocco	Mozam-bique	Nepal	Nether-lands
Front-line fighters						
F-14	-	-	-	-	-	-
F-15	-	-	-	-	-	-
F-16	-	-	-	-	-	181
F-18	-	-	-	-	-	-
F-117						
Harrier	-	-	-	-	-	-
MiG-29	-	-	-	-	-	-
MiG-31						
Mirage 2000	-	-	-	-	-	-
Su-24	-	-	-	-	-	-
Su-27						
Tornado	-	-	-	-	-	-
Capable fighters						
A-10A	-	-	-	-	-	-
A-37 Viggen	-	-	-	-	-	-
A-6	-	-	-	-	-	-
A-7	-	-	-	-	-	-
F-4	-	-	-	-	-	-
F-111	-	-	-	-	-	-
J-8						
Jaguar	-	-	-	-	-	-
MiG-23/27	-	-	-	-	-	-
MiG-25	-	-	-	-	-	-
Mirage F.1	-	-	29	-	-	-
Su-25	-	-	-	-	-	-
Super Entendard IV	-	-	-	-	-	-
Yak-38						
Old fighters						
A-4	-	-	-	-	-	-
Buccaneer	-	-	-	-	-	-
Crusader	-	-	-	-	-	-
F-5	11	-	18	-	-	-
F-104	-	-	-	-	-	-
F-35 Draken	-	-	-	-	-	-
F-86F						
G-91	-	-	-	-	-	-
Hunter	-	-	-	-	-	-
Kfir	-	-	-	-	-	-
MiG-15 (J-2)	-	2	-	-	-	-
MiG-17 (J-5)	-	-	-	-	-	-
MiG-19 (J-6)	-	-	-	-	-	-
MiG-21 (J-7)	-	15	-	43	-	-
Mirage III	-	-	-	-	-	-
Mirage 5	-	-	-	-	-	-
Ouragon						
Q-5	-	-	-	-	-	-
Su-7	-	-	-	-	-	-
Su-15						
Su-17/20/22	-	-	-	-	-	-
Super Mystere B2						

	Mexico	Mongolia	Morocco	Mozam-bique	Nepal	Nether-lands
Fighter summary						
Front-line	-	-	-	-	-	181
Less capable	-	-	29	-	-	-
Old	11	17	18	43	-	-
Total fighters	11	17	47	43	-	181
BOMBERS	-	-	-	-	-	-
SUPPORT AIRCRAFT						
Airlift	87	38	30	10	4	14
Armed trainers	12	-	46	-	-	-
Utility trainers	91	9	28	8	-	-
Counterinsurgency	80	-	-	-	-	-
MR/ASW	20	-	-	-	-	2
Tankers	-	-	4	-	-	-
Other	10	-	8	-	-	-
Total support aircraft	190	47	116	18	4	16

	New Zealand	Nicaragua	Niger	Nigeria	North Korea	Norway
FIGHTERS						
Front-line fighters						
F-14	-	-	-	-	-	-
F-15	-	-	-	-	-	-
F-16	-	-	-	-	-	61
F-18	-	-	-	-	-	-
F-117						
Harrier	-	-	-	-	-	-
MiG-29	-	-	-	-	30	-
MiG-31						
Mirage 2000	-	-	-	-	-	-
Su-24	-	-	-	-	-	-
Su-27						
Tornado	-	-	-	-	-	-
Capable fighters						
A-10A	-	-	-	-	-	-
A-37 Viggen	-	-	-	-	-	-
A-6	-	-	-	-	-	-
A-7	-	-	-	-	-	-
F-4	-	-	-	-	-	-
F-111	-	-	-	-	-	-
J-8						
Jaguar	-	-	-	15	-	-
MiG-23/27	-	-	-	-	46	-
MiG-25	-	-	-	-	-	-
Mirage F.1	-	-	-	-	-	-
Su-25	-	-	-	-	36	-
Super Entendard IV	-	-	-	-	-	-
Yak-38						
Old fighters						
A-4	21	-	-	-	-	-
Buccaneer	-	-	-	-	-	-
Crusader	-	-	-	-	-	-
F-5	-	-	-	-	-	29
F-104	-	-	-	-	-	-
F-35 Draken	-	-	-	-	-	-
F-86F						
G-91	-	-	-	-	-	-
Hunter	-	-	-	-	-	-
Kfir	-	-	-	-	-	-
MiG-15 (J-2)	-	-	-	-	50	-
MiG-17 (J-5)	-	-	-	-	350	-
MiG-19 (J-6)	-	-	-	-	190	-
MiG-21 (J-7)	-	-	-	22	170	-
Mirage III	-	-	-	-	-	-
Mirage 5	-	-	-	-	-	-
Ouragon						
Q-5	-	-	-	-	40	-
Su-7	-	-	-	-	20	-
Su-15						
Su-17/20/22	-	-	-	-	-	-
Super Mystere B2						

	New Zealand	Nicaragua	Niger	Nigeria	North Korea	Norway
Fighter summary						
Front-line	-	-	-	-	30	61
Less capable	-	-	-	15	82	-
Old	21	-	-	22	820	29
Total fighters	21	-	-	37	932	90
BOMBERS	-	-	-	-	-	-
SUPPORT AIRCRAFT						
Airlift	19	25	13	58	280	9
Armed trainers	14	-	-	56	-	-
Utility trainers	25	-	-	25	-	18
Counterinsurgency	-	16	-	23	-	-
MR/ASW	13	-	-	2	-	6
Tankers	-	-	-	-	-	-
Other	-	-	-	-	-	3
Total support aircraft	71	41	13	164	280	36

	Oman	Pakistan	Panama	Papua New Guinea	Paraguay	Peru
FIGHTERS						
Front-line fighters						
F-14	-	-	-	-	-	-
F-15	-	-	-	-	-	-
F-16	-	39	-	-	-	-
F-18	-	-	-	-	-	-
F-117						
Harrier	-	-	-	-	-	-
MiG-29	-	-	-	-	-	-
MiG-31						
Mirage 2000	-	-	-	-	-	12
Su-24	-	-	-	-	-	-
Su-27						
Tornado	-	-	-	-	-	-
Capable fighters						
A-10A	-	-	-	-	-	-
A-37 Viggen	-	-	-	-	-	-
A-6	-	-	-	-	-	-
A-7	-	-	-	-	-	-
F-4	-	-	-	-	-	-
F-111	-	-	-	-	-	-
J-8						
Jaguar	17	-	-	-	-	-
MiG-23/27	-	-	-	-	-	-
MiG-25	-	-	-	-	-	-
Mirage F.1	-	-	-	-	-	-
Su-25	-	-	-	-	-	-
Super Entendard IV	-	-	-	-	-	-
Yak-38						
Old fighters						
A-4	-	-	-	-	-	-
Buccaneer	-	-	-	-	-	-
Crusader	-	-	-	-	-	-
F-5	-	-	-	-	-	-
F-104	-	-	-	-	-	-
F-35 Draken	-	-	-	-	-	-
F-86F						
G-91	-	-	-	-	-	-
Hunter	-	-	-	-	-	-
Kfir	-	-	-	-	-	-
MiG-15 (J-2)	-	6	-	-	-	-
MiG-17 (J-5)	-	30	-	-	-	-
MiG-19 (J-6)	-	122	-	-	-	-
MiG-21 (J-7)	-	40	-	-	-	-
Mirage III	-	30	-	-	-	-
Mirage 5	-	58	-	-	-	16
Ouragon						
Q-5	-	50	-	-	-	-
Su-7	-	-	-	-	-	-
Su-15						
Su-17/20/22	-	-	-	-	-	41
Super Mystere B2						

	Oman	Pakistan	Panama	Papua New Guinea	Paraguay	Peru
Fighter summary						
Front-line	-	39	-	-	-	12
Less capable	17	-	-	-	-	-
Old	-	336	-	-	-	57
Total fighters	17	375	-	-	-	69
BOMBERS	-	-	-	-	-	15
SUPPORT AIRCRAFT						
Airlift	35	108	1	6	28	92
Armed trainers	16	63	-	-	-	53
Utility trainers	2	88	10	-	30	67
Counterinsurgency	7	-	-	-	5	-
MR/ASW	-	6	3	3	-	11
Tankers	-	-	-	-	-	1
Other	-	91	-	-	-	4
Total support aircraft	60	356	4	9	33	228

FIGHTERS	Philippines	Poland	Portugal	Qatar	Romania	Rwanda
Front-line fighters						
F-14	-	-	-	-	-	-
F-15	-	-	-	-	-	-
F-16	-	-	-	-	-	-
F-18	-	-	-	-	-	-
F-117						
Harrier	-	-	-	-	-	-
MiG-29	-	16	-	-	15	-
MiG-31						
Mirage 2000	-	-	-	-	-	-
Su-24	-	-	-	-	-	-
Su-27						
Tornado	-	-	-	-	-	-
Capable fighters						
A-10A	-	-	-	-	-	-
A-37 Viggen	-	-	-	-	-	-
A-6	-	-	-	-	-	-
A-7	-	-	38	-	-	-
F-4	-	-	-	-	-	-
F-111	-	-	-	-	-	-
J-8						
Jaguar	-	-	-	-	-	-
MiG-23/27	-	29	-	-	42	-
MiG-25	-	-	-	-	-	-
Mirage F.1	-	-	-	12	-	-
Su-25	-	-	-	-	-	-
Super Entendard IV	-	-	-	-	-	-
Yak-38						
Old fighters						
A-4	-	-	-	-	-	-
Buccaneer	-	-	-	-	-	-
Crusader	-	-	-	-	-	-
F-5	9	-	-	-	-	-
F-104	-	-	-	-	-	-
F-35 Draken	-	-	-	-	-	-
F-86F						
G-91	-	-	29	-	-	-
Hunter	-	-	-	-	-	-
Kfir	-	-	-	-	-	-
MiG-15 (J-2)	-	-	-	-	72	-
MiG-17 (J-5)	-	-	-	-	19	-
MiG-19 (J-6)	-	-	-	-	-	-
MiG-21 (J-7)	-	320	-	-	238	-
Mirage III	-	-	-	-	-	-
Mirage 5	-	-	-	-	-	-
Ouragon						
Q-5	-	-	-	-	-	-
Su-7	-	-	-	-	-	-
Su-15						
Su-17/20/22	-	125	-	-	-	-
Super Mystere B2						

	Philippines	Poland	Portugal	Qatar	Romania	Rwanda
Fighter summary						
Front-line	-	16	-	-	15	-
Less capable	-	29	38	12	42	-
Old	9	445	29	-	329	-
Total fighters	9	490	67	12	386	-
BOMBERS	-	-	-	-	-	-
SUPPORT AIRCRAFT						
Airlift	66	42	32	3	28	4
Armed trainers	11	-	52	6	150	-
Utility trainers	69	177	70	-	156	-
Counterinsurgency	8	-	-	-	-	2
MR/ASW	24	-	27	-	-	-
Tankers	-	-	-	-	-	-
Other	-	-	6	-	3	-
Total support aircraft	178	219	117	9	337	6

	Saudi Arabia	Senegal	Seychelles	Singapore	South Africa	South Korea
FIGHTERS						
Front-line fighters						
F-14	-	-	-	-	-	-
F-15	75	-	-	-	-	-
F-16	-	-	-	8	-	48
F-18	-	-	-	-	-	-
F-117						
Harrier	-	-	-	-	-	-
MiG-29	-	-	-	-	-	-
MiG-31						
Mirage 2000	-	-	-	-	-	-
Su-24	-	-	-	-	-	-
Su-27						
Tornado	52	-	-	-	-	-
Capable fighters						
A-10A	-	-	-	-	-	-
A-37 Viggen	-	-	-	-	-	-
A-6	-	-	-	-	-	-
A-7	-	-	-	-	-	-
F-4	-	-	-	-	-	150
F-111	-	-	-	-	-	-
J-8						
Jaguar	-	-	-	-	-	-
MiG-23/27	-	-	-	-	-	-
MiG-25	-	-	-	-	-	-
Mirage F.1	-	-	-	-	43	-
Su-25	-	-	-	-	-	-
Super Entendard IV	-	-	-	-	-	-
Yak-38						
Old fighters						
A-4	-	-	-	75	-	-
Buccaneer	-	-	-	-	-	-
Crusader	-	-	-	-	-	-
F-5	97	-	-	40	-	230
F-104	-	-	-	-	-	-
F-35 Draken	-	-	-	-	-	-
F-86F						
G-91	-	-	-	-	-	-
Hunter	-	-	-	24	-	-
Kfir	-	-	-	-	-	-
MiG-15 (J-2)	-	-	-	-	-	-
MiG-17 (J-5)	-	-	-	-	-	-
MiG-19 (J-6)	-	-	-	-	-	-
MiG-21 (J-7)	-	-	-	-	-	-
Mirage III	-	-	-	-	-	-
Mirage 5	-	-	-	-	-	-
Ouragon						
Q-5	-	-	-	-	-	-
Su-7	-	-	-	-	-	-
Su-15						
Su-17/20/22	-	-	-	-	-	-
Super Mystere B2						

	Saudi Arabia	Senegal	Seychelles	Singapore	South Africa	South Korea
Fighter summary						
Front-line	127	-	-	8	-	48
Less capable	-	-	-	-	43	150
Old	97	-	-	139	-	230
Total fighters	224	-	-	147	43	428
BOMBERS	-	-	-	-	-	-
SUPPORT AIRCRAFT						
Airlift	102	9	2	16	81	36
Armed trainers	41	5	-	-	216	87
Utility trainers	31	4	2	6-	271	59
Counterinsurgency	-	4	-	-	-	36
MR/ASW	-	1	1	-	8	24
Tankers	15	-	-	-	-	-
Other	5	-	-	4	4	-
Total support aircraft	194	23	5	20	580	242

	Former Soviet Union	Spain	Sri Lanka	Sudan	Suriname	Sweden
FIGHTERS						
Front-line fighters						
F-14	-	-	-	-	-	-
F-15	-	-	-	-	-	-
F-16	-	-	-	-	-	-
F-18	-	72	-	-	-	-
F-117						
Harrier	-	21	-	-	-	-
MiG-29	770	-	-	-	-	-
MiG-31	270					
Mirage 2000	-	-	-	-	-	-
Su-24	1130	-	-	-	-	-
Su-27	410					
Tornado	-	-	-	-	-	-
Capable fighters						
A-10A	-	-	-	-	-	-
A-37 Viggen	-	-	-	-	-	220
A-6	-	-	-	-	-	-
A-7	-	-	-	-	-	-
F-4	-	40	-	-	-	-
F-111	-	-	-	-	-	-
J-8						
Jaguar	-	-	-	-	-	-
MiG-23/27	2205	-	-	3	-	-
MiG-25	720	-	-	-	-	-
Mirage F.1	-	57	-	-	-	-
Su-25	535	-	-	-	-	-
Super Entendard IV	-	-	-	-	-	-
Yak-38	80					
Old fighters						
A-4	-	-	-	-	-	-
Buccaneer	-	-	-	-	-	-
Crusader	-	-	-	-	-	-
F-5	-	47	-	9	-	-
F-104	-	-	-	-	-	-
F-35 Draken	-	-	-	-	-	79
F-86F						
G-91	-	-	-	-	-	-
Hunter	-	-	-	-	-	-
Kfir	-	-	-	-	-	-
MiG-15 (J-2)	-	-	-	4	-	-
MiG-17 (J-5)	-	-	-	12	-	-
MiG-19 (J-6)	-	-	-	17	-	-
MiG-21 (J-7)	585	-	-	12	-	-
Mirage III	-	23	-	-	-	-
Mirage 5	-	-	-	-	-	-
Ouragon						
Q-5	-	-	-	-	-	-
Su-7	-	-	-	-	-	-
Su-15	500					
Su-17/20/22	810	-	-	-	-	-
Super Mystere B2						

	Former Soviet Union	Spain	Sri Lanka	Sudan	Suriname	Sweden
Fighter summary						
Front-line	2580	93	-	-	-	-
Less capable	3540	97	-	3	-	220
Old	1895	70	-	54	-	79
Total fighters	8015	260	-	57	-	299
BOMBERS	1043	-	-	-	-	-
SUPPORT AIRCRAFT						
Airlift	590	121	15	23	1	76
Armed trainers	600	-	-	6	-	-
Utility trainers	60	180	12	-	-	7-Feb
Counterinsurgency	-	-	11	-	5	-
MR/ASW	3-Mar	20	6	2	-	25
Tankers	79	8	-	-	-	-
Other	285	37	-	-	-	19
Total support aircraft	34488	166	44	31	6	33030

	Switzerland	Syria	Taiwan	Tanzania	Thailand	Togo
FIGHTERS						
Front-line fighters						
F-14	-	-	-	-	-	-
F-15	-	-	-	-	-	-
F-16	-	-	-	-	18	-
F-18	-	-	-	-		-
F-117						
Harrier	-	-	-	-	-	-
MiG-29	-	30	-	-	-	-
MiG-31						
Mirage 2000	-	-	-	-	-	-
Su-24	-	22	-	-	-	-
Su-27						
Tornado	-	-	-	-	-	-
Capable fighters						
A-10A	-	-	-	-	-	-
A-37 Viggen	-	-	-	-	-	-
A-6	-	-	-	-	-	-
A-7	-	-	-	-	-	-
F-4	-	-	-	-	-	-
F-111	-	-	-	-	-	-
J-8						
Jaguar	-	-	-	-	-	-
MiG-23/27	-	156	-	-	-	-
MiG-25	-	41	-	-	-	-
Mirage F.1	-	-	-	-	-	-
Su-25	-	-	-	-	-	-
Super Entendard IV	-	-	-	-	-	-
Yak-38						
Old fighters						
A-4	-	-	-	-	-	-
Buccaneer	-	-	-	-	-	-
Crusader	-	-	-	-	-	-
F-5	104	-	278	-	60	-
F-104	-	-	149	-	-	-
F-35 Draken	-	-	-	-	-	-
F-86F						
G-91	-	-	-	-	-	-
Hunter	126	-	-	-	-	-
Kfir	-	-	-	-	-	-
MiG-15 (J-2)	-	-	-	2	-	-
MiG-17 (J-5)	-	-	-	-	-	-
MiG-19 (J-6)	-	-	-	10	-	-
MiG-21 (J-7)	-	222	-	11	-	-
Mirage III	52	-	-	-	-	-
Mirage 5	-	-	-	-	-	-
Ouragon						
Q-5	-	-	-	-	-	-
Su-7	-	-	-	-	-	-
Su-15						
Su-17/20/22	-	90	-	-	-	-
Super Mystere B2						

	Switzerland	Syria	Taiwan	Tanzania	Thailand	Togo
Fighter summary						
Front-line	-	52	-	-	18	-
Less capable	-	197	-	-	-	-
Old	282	312	427	23	60	-
Total fighters	282	561	427	23	78	-
BOMBERS	-	-	-	-	-	-
SUPPORT AIRCRAFT						
Airlift	20	28	81	17	98	6
Armed trainers	20	100	30	-	47	13
Utility trainers	117	10	2-Jan	5	116	-
Counterinsurgency	-	-	60	-	181	11
MR/ASW	-	-	40	-	23	-
Tankers	-	-	-	-	-	-
Other	-	-	-	-	8	-
Total support aircraft	157	128	32985	22	473	30

	Transkei (S. Africa)	Trinidad and Tobago	Tunisia	Turkey	UAE	Uganda
FIGHTERS						
Front-line fighters						
F-14	-	-	-	-	-	-
F-15	-	-	-	-	-	-
F-16	-	-	-	71	-	-
F-18	-	-	-	-	-	-
F-117						
Harrier	-	-	-	-	-	-
MiG-29	-	-	-	-	-	-
MiG-31						
Mirage 2000	-	-	-	-	36	-
Su-24	-	-	-	-	-	-
Su-27						
Tornado	-	-	-	-	-	-
Capable fighters						
A-10A	-	-	-	-	-	-
A-37 Viggen	-	-	-	-	-	-
A-6	-	-	-	-	-	-
A-7	-	-	-	-	-	-
F-4	-	-	-	171	-	-
F-111	-	-	-	-	-	-
J-8						
Jaguar	-	-	-	-	-	-
MiG-23/27	-	-	-	-	-	-
MiG-25	-	-	-	-	-	-
Mirage F.1	-	-	-	-	-	-
Su-25	-	-	-	-	-	-
Super Entendard IV	-	-	-	-	-	-
Yak-38						
Old fighters						
A-4	-	-	-	-	-	-
Buccaneer	-	-	-	-	-	-
Crusader	-	-	-	-	-	-
F-5	-	-	30	180	-	-
F-104	-	-	-	250	-	-
F-35 Draken	-	-	-	-	-	-
F-86F						
G-91	-	-	-	-	-	-
Hunter	-	-	-	-	-	-
Kfir	-	-	-	-	-	-
MiG-15 (J-2)	-	-	-	-	-	-
MiG-17 (J-5)	-	-	-	-	-	8
MiG-19 (J-6)	-	-	-	-	-	-
MiG-21 (J-7)	-	-	-	-	-	5
Mirage III	-	-	-	-	14	-
Mirage 5	-	-	-	-	15	-
Ouragon						
Q-5	-	-	-	-	-	-
Su-7	-	-	-	-	-	-
Su-15						
Su-17/20/22	-	-	-	-	-	-
Super Mystere B2						

	Transkei (S. Africa)	Trinidad and Tobago	Tunisia	Turkey	UAE	Uganda
Fighter summary						
Front-line	-	-	-	71	36	-
Less capable	-	-	-	171	-	-
Old	-	-	30	430	29	13
Total fighters	-	-	30	672	65	13
BOMBERS	-	-	-	-	-	-
SUPPORT AIRCRAFT						
Airlift	2	3	4	96	8	3
Armed trainers	-	-	19	165	35	5
Utility trainers	-	-	21	159	30	22
Counterinsurgency	-	-	-	-	-	-
MR/ASW	-	-	-	20	-	-
Tankers	-	-	-	-	-	-
Other	-	-	-	-	4	-
Total support aircraft	2	3	44	420	47	30

	UK	United States	Uruguay	Venezuela	Vietnam	Yemen
FIGHTERS						
Front-line fighters						
F-14	-	514	-	-	-	-
F-15	-	26-Jan	-	-	-	-
F-16	-	1682	-	24	-	-
F-18	-	815	-	-	-	-
F-117		59				
Harrier	101	193	-	-	-	-
MiG-29	-	-	-	-	-	-
MiG-31						
Mirage 2000	-	-	-	-	-	-
Su-24	-	-	-	-	-	-
Su-27						
Tornado	310	-	-	-	-	-
Capable fighters						
A-10A	-	641	-	-	-	-
A-37 Viggen	-	-	-	-	-	-
A-6	-	331	-	-	-	-
A-7	-	330	-	-	-	-
F-4	-	1027	-	-	-	-
F-111	-	294	-	-	-	-
J-8						
Jaguar	115	-	-	-	-	-
MiG-23/27	-	4	-	-	-	-
MiG-25	-	-	-	-	-	-
Mirage F.1	-	-	-	-	-	-
Su-25	-	-	-	-	-	-
Super Entendard IV	-	-	-	-	-	-
Yak-38						
Old fighters						
A-4	-	326	-	-	-	-
Buccaneer	62	-	-	-	-	-
Crusader	-	-	-	-	-	-
F-5	-	46	-	27	-	14
F-104	-	-	-	-	-	-
F-35 Draken	-	-	-	-	-	-
F-86F						
G-91	-	-	-	-	-	-
Hunter	26	-	-	-	-	-
Kfir	-	-	-	-	-	-
MiG-15 (J-2)	-	-	-	-	-	-
MiG-17 (J-5)	-	-	-	-	-	-
MiG-19 (J-6)	-	-	-	-	-	-
MiG-21 (J-7)	-	24	-	-	125	50
Mirage III	-	-	-	4	-	-
Mirage 5	-	-	-	9	-	-
Ouragon						
Q-5	-	-	-	-	-	-
Su-7	-	-	-	-	-	-
Su-15						
Su-17/20/22	-	-	-	-	60	37
Super Mystere B2						

	UK	United States	Uruguay	Venezuela	Vietnam	Yemen
Fighter summary						
Front-line	411	4289	-	24	-	-
Less capable	115	2627	-	-	-	-
Old	88	396	-	40	185	101
Total fighters	614	7312	-	64	185	101
BOMBERS	-	307	-	-	-	-
SUPPORT AIRCRAFT						
Airlift	2	1198	22	115	68	27
Armed trainers	144	1457	25	19	-	-
Utility trainers	54	1837	46	38	-	-
Counterinsurgency	-	127	12	23	-	-
MR/ASW	-	434	16	5	4	-
Tankers	-	749	-	-	-	-
Other	20	421	2	-	2	-
Total support aircraft	200	6223	123	200	74	27

	Yugoslavia	Zaire	Zambia	Zimbabwe
FIGHTERS				
Front-line fighters				
F-14	-	-	-	-
F-15	-	-	-	-
F-16	-	-	-	-
F-18	-	-	-	-
F-117				
Harrier	-	-	-	-
MiG-29	26	-	-	-
MiG-31				
Mirage 2000	-	-	-	-
Su-24	-	-	-	-
Su-27				
Tornado	-	-	-	-
Capable fighters				
A-10A	-	-	-	-
A-37 Viggen	-	-	-	-
A-6	-	-	-	-
A-7	-	-	-	-
F-4	-	-	-	-
F-111	-	-	-	-
J-8				
Jaguar	-	-	-	-
MiG-23/27	-	-	-	-
MiG-25	-	-	-	-
Mirage F.1	-	-	-	-
Su-25	-	-	-	-
Super Entendard IV	-	-	-	-
Yak-38				
Old fighters				
A-4	-	-	-	-
Buccaneer	-	-	-	-
Crusader	-	-	-	-
F-5	-	-	-	-
F-104	-	-	-	-
F-35 Draken	-	-	-	-
F-86F				
G-91	-	-	-	-
Hunter	-	-	-	10
Kfir	-	-	-	-
MiG-15 (J-2)	-	-	-	-
MiG-17 (J-5)	-	-	-	-
MiG-19 (J-6)	-	-	22	-
MiG-21 (J-7)	115	-	14	48
Mirage III	-	-	-	-
Mirage 5	-	8	-	-
Ouragon				
Q-5	-	-	-	-
Su-7	-	-	-	-
Su-15				
Su-17/20/22	-	-	-	-
Super Mystere B2				

	Yugoslavia	Zaire	Zambia	Zimbabwe
Fighter summary				
Front-line	26	-	-	-
Less capable	-	-	-	-
Old	115	8	36	58
Total fighters	141	8	36	58
BOMBERS	-	-	-	-
SUPPORT AIRCRAFT				
Airlift	88	23	18	43
Armed trainers	328	20	27	6
Utility trainers	170	24	-	1
Counterinsurgency	-	-	37	16
MR/ASW	-	-	-	-
Tankers	-	-	-	-
Other	20	-	-	-
Total support aircraft	416	67	82	66

4. Details on the Economic Data

Economic Data

The potential for nations to modernize their air forces will depend heavily on the current and future health of their economies, which in turn provides the funds needed by governments for discretionary spending on items such as national defense. This study examines recent historical patterns of global wealth distribution, central government expenditures, and military expenditures.

Economic data were obtained from *World Military Expenditures and Arms Transfers 1990 (WMEAT)*, a publication of the U.S. Arms Control and Disarmament Agency (ACDA). This publication contains data from 1979 to 1989. Although more recent data for the period including the collapse of the Warsaw Pact would have been highly desirable, such data simply were not available on a global basis, but estimates for various key nations were developed. The International Monetary Fund's *International Financial Statistics* and other data sources helped to make informed judgments where *WMEAT* data were missing.

Purchasing power parity (PPP) exchange rates were employed for estimating the economic data of the former major centrally planned economies—the former Soviet Union and other Warsaw Pact countries and China. Compilation of economic data of the former Soviet Union, other former Warsaw Pact countries, and China involves procedures that have considerable uncertainty understandably associated with them. Consequently, the resulting data for these countries should be regarded cautiously.

GNP Data

The ACDA GNP estimates for most noncommunist countries derive from the International Bank for Reconstruction and Development (World Bank). ACDA data for former communist countries come from the Central Intelligence Agency's (CIA) *Handbook of Economic Statistics 1990*. GNP is the measure of the aggregate value of a nation's output of goods and services. Data are occasionally available for gross domestic product (GDP) only. GDP is defined as GNP less net factor income/payments abroad. This difference is of little significance for our purposes, as GNP and GDP normally differ only by very small amounts.

Central Government Expenditures Data

The principal sources for ACDA data on central government expenditures are the IMF's *Government Finance Statistics Yearbook*, the IMF's *International Financial Statistics*, the Organization for Economic Cooperation and Development's (OECD) *Economic Surveys*, and the CIA's *World Factbook*.

Military Expenditures Data

NATO publications were the principal sources of ACDA data on NATO country military expenditures, which are defined as non-civilian-type defense ministry expenditures, including foreign military assistance. For other noncommunist countries, the defense ministry expenditures were obtained from the IMF's *Government Finance Statistics Yearbook*. Data for the former Soviet Union are based on the CIA's estimate of how much American resources would be required to build the Soviet military in the United States. China data are from the Defense Intelligence Agency. ACDA obtained military expenditure data on the former Warsaw Pact countries from the Congressional Joint Economic Committee.

Cost Comparisons

WMEAT contains historical country-by-country data on a range of key economic and military manpower variables. For analytic purposes, measuring the variables in constant-year currency is critical. Furthermore, expressing all economic data in a single currency is essential for making international comparisons. ACDA deflates each country's economic time-series to a base year constant currency by using each country's GNP deflator. In order to convert every currency time-series to a single currency (the U.S. dollar), ACDA employed exchange rates from the IMF's *International Financial Statistics* for the base year chosen. Using base year currency exchange rates avoids the distorting effect of fluctuating exchange rates that would result if one were to use different exchange rates for each year.

In the current international financial system, currency exchange rates fluctuate for a variety of reasons that often have little to do with a country's economic vitality, but much to do with supply and demand on the part of international currency traders for currencies. For instance, a periodic rise or fall in a country's interest rates will cause changes in the exchange rate for its currency. GNP estimates that use such exchange rates can yield different values from day to day as a result of these fluctuations, even though the economic strength of the country remains unchanged. PPP exchange rates circumvent this problem. A PPP exchange rate is that rate which renders equivalent the purchasing power of currencies across countries. Accordingly, PPP exchange rates were employed for estimating the economic data of the former major centrally planned economies—the former Soviet Union, the other Warsaw Pact countries, and China.

5. Global Gross National Products, 1979–1989

Figures are in millions of FY89 dollars. Where estimates are employed, figures are denoted in italics. Estimates were typically interpolated between two known values.

Country	GNP (US$ M 1989)					
	1979	1980	1981	1982	1983	1984
Afghanistan	4,152	4,024	4,135	3,955	3,821	3,609
Albania	*3,074*	*3,074*	*3,074*	*3,074*	*3,074*	*3,074*
Algeria	34,800	35,270	36,300	38,620	40,880	43,130
Angola	3,292	3,477	3,406	3,591	3,693	4,454
Argentina	64,230	64,630	58,300	54,410	55,600	57,060
Australia	200,500	205,400	211,200	210,200	210,300	224,700
Austria	102,300	105,300	105,000	106,200	108,300	109,900
Bahamas	no data					
Bahrain	3,584	4,357	4,609	4,478	4,347	4,401
Bangladesh	14,460	14,690	15,650	15,690	16,230	16,990
Belgium	127,000	131,800	130,500	131,600	132,100	135,000
Belize	no data					
Benin	1,272	1,354	1,478	1,587	1,540	1,570
Bolivia	4,568	4,446	4,460	4,217	3,961	3,976
Botswana	764	888	1,011	1,010	1,217	1,399
Brazil	353,400	384,000	363,300	360,900	346,600	364,900
Brunei	no data					
Bulgaria	46,400	45,040	46,270	47,760	46,870	48,380
Burkina Faso	1,668	1,686	1,753	1,917	1,923	1,894
Burundi	698	709	795	766	821	818
Cambodia	*807*	*807*	807	*807*	*807*	*807*
Cameroon	7,217	8,185	9,244	9,770	10,540	11,130
Canada	392,200	398,000	409,800	395,900	410,100	435,700
Cape Verde	120	172	181	188	204	212
Central African Republic	1,014	976	958	1,008	939	1,021
Chad	635	600	609	641	739	747
Chile	17,240	18,520	19,250	16,060	15,950	16,570
China	258,100	274,600	288,000	312,000	343,100	389,700
Colombia	27,410	28,620	29,110	29,120	29,460	30,160
Congo	1,135	1,299	1,572	1,902	2,001	2,099
Costa Rica	4,151	4,129	3,947	3,511	3,694	4,038
Cote d'Ivoire	8,468	9,259	9,501	9,461	9,143	8,989
Cuba	26,690	25,980	28,010	28,760	30,680	31,640
Cyprus	2,567	2,721	2,782	2,947	3,074	3,351
Czech & Slovak Republics	102,500	104,900	104,500	106,400	108,000	110,600
Denmark	86,550	85,690	84,420	86,320	88,700	92,000
Djibouti	no data					
Dominican Republic	5,056	5,078	5,130	5,426	5,658	5,733
Ecuador	7,622	7,937	8,225	8,131	7,895	8,056
Egypt	41,350	45,100	46,400	51,320	55,260	58,570
El Salvador	7,258	6,580	5,975	5,591	5,631	5,761
Equatorial Guinea	108	*108*	*108*	*108*	*108*	*108*
Ethiopia	4,704	4,924	5,012	5,090	5,345	5,220
Fiji	1,095	1,077	1,153	1,062	1,015	1,101
Finland	79,350	83,580	84,630	87,510	90,170	92,730
Former Soviet Union	2,235,000	2,257,000	2,284,000	2,337,000	2,390,000	2,419,000
France	776,600	790,000	798,300	816,400	819,300	828,000
Gabon	3,034	3,137	3,018	3,112	3,157	3,093
Germany (FRG)	995,500	1,010,000	1,010,000	1,001,000	1,020,000	1,053,000
Germany (GDR)	129,400	132,000	134,800	134,200	136,900	140,700
Ghana	4,255	4,251	4,110	3,825	3,657	3,956
Greece	46,880	47,770	47,690	47,650	47,390	48,340
Guatemala	7,588	7,822	7,846	7,553	7,366	7,332
Guinea	2,546	2,592	2,569	2,573	2,581	2,609

Country	GNP (US$ M 1989)				
	1985	1986	1987	1988	1989
Afghanistan	3,322	3,079	3,079	3,079	3,079
Albania	3,074	3,107	3,162	3,214	3,800
Algeria	45,550	46,090	45,800	44,130	45,290
Angola	4,575	5,227	6,003	6,101	6,031
Argentina	55,190	59,210	60,490	58,780	54,080
Australia	236,200	241,100	250,300	257,800	270,800
Austria	112,700	113,700	115,800	120,600	125,200
Bahamas					
Bahrain	3,886	3,203	3,102	3,032	2,999
Bangladesh	17,600	18,430	19,150	19,550	20,020
Belgium	135,900	138,600	141,700	147,600	154,600
Belize					
Benin	1,652	1,638	1,575	1,615	1,637
Bolivia	3,918	3,835	3,934	4,090	4,226
Botswana	1,448	1,637	1,749	1,920	2,188
Brazil	396,900	431,100	450,700	448,200	462,300
Brunei					
Bulgaria	46,920	48,180	48,260	49,940	49,590
Burkina Faso	2,090	2,273	2,319	2,537	2,552
Burundi	914	933	991	1,043	1,058
Cambodia	807	807	807	807	807
Cameroon	11,980	13,060	12,600	11,720	11,100
Canada	456,900	469,500	492,000	516,800	531,000
Cape Verde	231	240	257	268	282
Central African Republic	1,057	1,062	1,030	1,046	1,088
Chad	916	875	845	993	1,002
Chile	17,350	18,340	19,790	21,050	23,300
China	439,800	474,600	523,100	582,000	603,500
Colombia	31,020	32,440	34,170	35,440	36,890
Congo	2,070	1,975	1,983	1,978	2,008
Costa Rica	4,095	4,314	4,419	4,615	4,899
Cote d'Ivoire	9,205	9,791	9,644	8,728	8,728
Cuba	33,610	34,870	36,250	36,140	35,460
Cyprus	3,515	3,643	3,896	4,229	4,468
Czech & Slovak Republics	111,300	113,600	114,600	121,900	123,200
Denmark	95,910	99,470	99,010	98,940	100,400
Djibouti					
Dominican Republic	5,442	5,634	5,963	5,970	6,422
Ecuador	8,575	8,823	8,280	9,644	9,668
Egypt	60,380	61,310	63,970	67,580	69,780
El Salvador	5,876	5,908	6,073	6,263	6,335
Equatorial Guinea	109	110	115	124	125
Ethiopia	4,842	5,172	5,641	5,825	5,959
Fiji	1,054	1,151	1,059	1,036	1,164
Finland	96,130	98,720	102,000	107,400	112,800
Former Soviet Union	2,442,000	2,525,000	2,574,000	2,630,000	2,664,000
France	843,500	864,700	885,800	920,500	954,100
Gabon	3,297	3,674	2,997	3,000	3,119
Germany (FRG)	1,073,000	1,098,000	1,117,000	1,158,000	1,207,000
Germany (GDR)	145,000	147,200	149,600	157,500	159,400
Ghana	4,143	4,342	4,548	4,839	5,134
Greece	49,440	49,520	49,590	51,680	52,930
Guatemala	7,349	7,296	7,558	7,890	8,208
Guinea	2,740	2,394	2,459	2,558	2,550

Country	GNP (US$ M 1989)					
	1979	1980	1981	1982	1983	1984
Guinea-Bissau	126	107	125	130	128	135
Guyana	475	477	471	406	371	354
Haiti	2,146	2,309	2,248	2,168	2,176	2,192
Honduras	3,586	3,612	3,696	3,561	3,605	3,696
Hungary	57,390	57,930	58,380	60,460	59,850	61,380
India	155,600	165,500	175,900	181,700	194,700	201,600
Indonesia	47,270	51,720	56,540	57,990	63,010	67,550
Iran	68,920	55,620	60,360	68,440	75,780	78,060
Iraq	68,120	81,880	49,150	51,540	48,290	49,660
Ireland	24,770	25,430	26,100	25,910	25,500	26,090
Israel	31,620	34,010	36,120	36,790	38,090	37,630
Italy	666,672	692,070	695,334	695,844	703,086	725,526
Jamaica	2,932	2,741	2,861	2,909	3,018	2,796
Japan	2,034,900	2,121,600	2,199,120	2,267,460	2,340,900	2,459,220
Jordan	3,023	3,559	3,798	4,084	4,059	4,266
Kenya	5,626	5,988	6,230	6,455	6,528	6,610
Kuwait	42,210	45,400	41,740	34,060	31,400	31,330
Laos	*581*	*581*	*581*	581	*600*	615
Lebanon	9,003	9,633	9,536	8,106	6,819	5,006
Lesotho	611	610	630	701	685	715
Liberia	1,445	1,376	1,350	1,267	1,207	1,160
Libya	40,810	51,730	42,140	35,980	33,760	28,900
Luxembourg	5,889	6,068	6,262	6,956	7,295	7,601
Madagascar	2,423	2,423	2,158	2,112	2,122	2,119
Malawi	1,182	1,141	1,080	1,116	1,168	1,238
Malaysia	21,420	23,297	25,041	26,194	27,203	29,019
Mali	1,583	1,559	1,611	1,721	1,626	1,630
Malta	1,276	1,394	1,533	1,597	1,649	1,608
Mauritania	837	892	941	891	946	894
Mexico	157,100	168,100	181,400	176,400	167,600	174,700
Mongolia	*2,000*	*2,000*	*2,000*	*2,000*	*2,000*	*2,000*
Morocco	14,440	15,690	15,030	16,550	16,420	17,130
Mozambique	1,343	1,376	1,375	1,301	1,122	1,161
Myanmar/Burma	13,320	14,370	15,290	16,130	16,700	17,450
Namibia	no data					
Nepal	1,886	1,849	2,004	2,076	2,016	2,201
Netherlands	189,100	191,400	189,800	187,400	190,400	195,500
New Zealand	33,120	33,600	34,700	35,170	35,340	36,540
Nicaragua	1,208	1,269	1,352	1,337	1,410	1,331
Niger	2,100	2,257	2,286	2,202	2,147	1,823
Nigeria	27,510	27,810	26,640	26,920	25,560	23,600
North Korea	32,905	31,600	29,886	29,070	29,060	29,050
North Yemen	4,234	4,425	4,701	5,154	5,216	5,328
Norway	62,150	65,040	65,640	65,690	68,950	73,120
Oman	4,817	7,065	7,934	7,827	7,685	8,222
Pakistan	19,140	21,090	22,840	24,300	25,840	27,160
Panama	3,857	4,256	4,467	4,665	4,651	4,682
Papua New Guinea	3,005	2,903	2,904	2,873	2,942	3,005
Paraguay	2,985	3,439	3,747	3,682	3,416	3,507
Peru	40,720	43,180	45,450	45,620	39,250	40,970
Philippines	36,350	38,170	39,440	40,190	40,680	37,750
Poland	163,500	159,300	150,900	149,500	156,900	162,700
Portugal	33,640	35,030	34,730	35,390	35,330	34,430
Qatar	8,611	12,390	12,360	10,290	8,423	8,487
Romania	110,300	110,500	109,900	109,900	109,000	113,300

Country	GNP (US$ M 1989)				
	1985	1986	1987	1988	1989
Guinea-Bissau	142	137	142	142	142
Guyana	350	342	312	316	316
Haiti	2,204	2,218	2,230	2,196	2,196
Honduras	3,797	3,886	4,058	4,184	4,184
Hungary	59,850	61,180	62,220	65,930	64,740
India	214,200	224,100	234,100	256,300	267,400
Indonesia	69,520	74,040	78,110	83,080	89,370
Iran	80,880	74,110	73,010	74,260	77,540
Iraq	49,660	38,760	38,490	41,250	41,250
Ireland	26,310	26,010	27,460	27,790	28,900
Israel	39,130	41,020	43,780	44,700	44,890
Italy	746,334	768,570	792,846	812,430	812,430
Jamaica	2,564	2,630	2,765	2,793	2,793
Japan	2,580,600	2,642,820	2,756,040	2,913,120	#######
Jordan	4,366	4,477	4,403	4,207	4,207
Kenya	6,901	7,388	7,803	8,254	8,254
Kuwait	28,310	27,110	28,040	26,800	31,880
Laos	615	615	615	615	615
Lebanon	2,953	2,993	1,915	1,915	1,915
Lesotho	694	695	714	739	739
Liberia	1,207	1,176	1,164	1,164	1,164
Libya	30,740	24,500	24,930	31,330	22,200
Luxembourg	7,898	8,038	8,051	8,454	8,879
Madagascar	2,153	2,186	2,168	2,253	2,340
Malawi	1,299	1,303	1,321	1,388	1,388
Malaysia	28,682	29,427	30,855	32,977	32,977
Mali	1,606	1,909	1,931	1,912	1,912
Malta	1,638	1,661	1,700	1,776	1,847
Mauritania	911	955	839	882	943
Mexico	181,200	172,500	177,100	181,300	186,700
Mongolia	2,000	2,000	2,000	2,000	2,000
Morocco	17,950	19,800	19,260	21,080	21,740
Mozambique	1,046	1,045	988	1,000	1,106
Myanmar/Burma	17,930	17,730	17,080	15,190	16,330
Namibia					
Nepal	2,334	2,432	2,512	2,765	2,811
Netherlands	200,800	205,400	207,800	213,400	222,500
New Zealand	38,640	38,060	37,880	38,710	39,090
Nicaragua	1,268	1,145	1,272	1,139	1,106
Niger	1,923	2,056	1,955	2,050	1,987
Nigeria	25,450	26,590	25,320	26,240	27,520
North Korea	29,325	29,560	29,713	29,784	29,784
North Yemen	5,564	5,911	6,123	6,543	6,776
Norway	77,810	81,290	84,230	84,570	88,510
Oman	9,051	7,099	7,845	7,472	7,635
Pakistan	29,100	30,610	32,520	35,060	36,810
Panama	4,856	5,051	5,167	4,215	4,134
Papua New Guinea	3,112	3,238	3,293	3,413	3,375
Paraguay	3,620	3,575	3,709	3,967	4,323
Peru	41,470	46,430	50,380	45,050	40,820
Philippines	36,220	36,890	39,080	41,650	43,960
Poland	164,300	169,600	167,100	177,900	174,700
Portugal	35,820	38,050	40,350	42,180	44,620
Qatar	7,652	6,300	6,371	6,277	6,870
Romania	113,100	114,900	112,500	117,000	113,400

Country	GNP (US$ M 1989)					
	1979	1980	1981	1982	1983	1984
Rwanda	1,687	1,869	2,070	2,078	2,193	2,102
Senegal	3,565	3,446	3,388	3,899	4,013	3,782
Seychelles	no data					
Sierra Leone	630	658	703	712	706	716
Singapore	14,100	14,990	16,370	17,680	19,580	21,720
Somali Republic	1,046	1,001	1,052	1,085	997	984
South Africa	69,180	73,080	80,510	77,290	76,610	80,770
South Korea	89,168	84,935	89,964	97,022	107,508	116,688
South Yemen	992	949	1,014	1,238	1,183	1,661
Spain	286,300	289,100	286,900	290,200	295,000	300,400
Sri Lanka	4,648	4,911	5,110	5,497	5,684	5,832
Sudan	13,170	13,260	13,300	14,830	15,060	14,070
Suriname	1,626	1,564	1,724	1,613	1,446	1,381
Sweden	154,700	156,100	155,000	155,800	158,300	164,300
Switzerland	146,600	152,800	156,300	154,500	156,500	160,900
Syria	14,820	16,550	18,090	18,320	18,610	17,890
Taiwan	68,390	73,250	77,470	80,610	87,580	97,730
Tanzania	2,268	2,327	2,300	2,292	2,277	2,371
Thailand	34,180	36,000	38,050	39,600	42,790	45,630
The Gambia	165	149	162	174	171	173
Togo	1,077	1,225	1,176	1,125	1,058	1,067
Trinidad and Tobago	5,307	5,884	6,420	6,809	6,103	5,458
Tunisia	6,715	7,259	7,616	7,565	7,945	8,419
Turkey	50,910	50,310	52,240	54,700	56,690	59,930
UAE	38,030	46,170	46,800	41,460	35,690	33,870
Uganda	3,081	2,958	3,188	3,201	3,278	3,079
United Kingdom	672,400	652,500	645,000	656,200	681,200	695,500
United States	4,032,000	4,025,000	4,103,000	3,999,000	4,142,000	4,422,000
Uruguay	7,891	8,343	8,528	7,602	6,942	6,730
Venezuela	44,920	43,100	43,070	41,110	39,320	39,500
Vietnam	12,380	12,380	13,000	13,000	13,000	13,000
Yugoslavia	56,430	57,410	57,860	57,740	57,020	57,960
Zaire	7,985	8,117	8,405	8,295	8,410	8,333
Zambia	3,940	4,144	4,632	4,382	4,210	4,087
Zimbabwe	3,776	4,197	4,663	4,728	4,794	4,748

Country	GNP (US$ M 1989)				
	1985	1986	1987	1988	1989
Rwanda	2,195	2,313	2,305	2,309	2,160
Senegal	3,908	4,101	4,271	4,495	4,441
Seychelles					
Sierra Leone	694	605	699	699	686
Singapore	21,720	22,120	23,680	26,480	28,890
Somali Republic	1,063	1,093	1,178	1,156	1,173
South Africa	80,260	80,530	82,930	87,040	86,750
South Korea	123,828	137,904	154,530	172,278	172,278
South Yemen	1,572	1,321	1,328	1,306	1,273
Spain	308,500	319,300	337,500	354,200	370,700
Sri Lanka	6,280	6,551	6,626	6,786	6,939
Sudan	13,310	14,720	15,000	14,200	15,640
Suriname	1,348	1,352	1,198	1,244	1,319
Sweden	168,000	172,300	178,300	182,000	185,800
Switzerland	166,700	169,800	172,900	179,400	184,300
Syria	18,980	17,980	17,940	20,260	19,320
Taiwan	103,200	116,100	129,900	140,100	150,200
Tanzania	2,366	2,460	2,453	2,562	2,642
Thailand	46,990	48,940	53,130	61,080	68,770
The Gambia	162	164	183	197	210
Togo	1,130	1,171	1,186	1,249	1,292
Trinidad and Tobago	4,886	4,860	4,352	4,000	3,735
Tunisia	8,793	8,657	9,098	9,237	9,616
Turkey	63,240	68,250	73,450	76,050	77,280
UAE	32,470	25,190	26,630	25,590	27,760
Uganda	3,215	3,279	3,549	3,796	4,045
United Kingdom	720,200	746,900	779,200	815,400	834,400
United States	4,571,000	4,696,000	4,857,000	5,073,000	5,201,000
Uruguay	6,760	7,457	7,953	7,975	8,066
Venezuela	39,550	42,570	43,900	46,570	41,460
Vietnam	13,000	13,760	13,550	13,740	14,200
Yugoslavia	58,230	60,750	59,810	58,180	58,640
Zaire	8,578	8,976	8,975	9,310	9,152
Zambia	4,126	3,839	4,003	4,507	4,655
Zimbabwe	5,021	5,142	5,105	5,491	5,742

6. Global Central Government Expenditures, 1979–1989

Figures are in millions of FY89 dollars. Where estimates are employed, figures are denoted in italics. Estimates were typically interpolated between two known values.

Country	Central Government Expenditures (US$ M 1989)					
	1979	1980	1981	1982	1983	1984
Afghanistan	551	762	780	587	391	508
Albania	1,570	1,570	1,570	1,520	1,520	1,524
Algeria	9,508	10,070	11,480	13,890	15,720	17,440
Angola	2,729	2,729	2,729	2,729	2,264	2,700
Argentina	12,360	13,570	14,980	13,620	17,180	12,240
Australia	52,660	51,850	54,110	55,430	60,770	65,980
Austria	39,700	40,670	41,660	42,730	44,520	44,680
Bahamas	no data					
Bahrain	1,207	1,480	1,523	1,754	1,898	1,703
Bangladesh	1,947	1,712	2,518	2,272	2,002	2,023
Belgium	64,130	67,340	73,290	74,140	75,280	79,320
Belize	no data					
Benin	290	277	309	402	400	398
Bolivia	630	679	639	1,009	460	1,588
Botswana	360	411	437	527	553	748
Brazil	83,900	99,090	102,500	104,700	111,700	108,500
Brunei	no data					
Bulgaria	17,320	16,930	19,470	20,160	19,470	20,000
Burkina Faso	209	198	220	229	193	225
Burundi	163	155	196	192	193	112
Cambodia	269	269	269	269	269	269
Cameroon	1,179	1,282	1,915	2,018	2,576	2,700
Canada	85,720	89,900	95,960	102,900	106,600	115,300
Cape Verde	107	107	147	200	200	200
Central African Republic	187	176	212	230	230	251
Chad	87	65	65	65	46	44
Chile	3,758	3,853	4,329	3,908	3,829	4,134
China	82,090	74,420	67,210	69,330	76,490	86,740
Colombia	3,438	3,952	4,277	4,752	4,688	4,778
Congo	713	713	673	1,036	848	994
Costa Rica	1,079	1,093	922	772	981	1,017
Cote d'Ivoire	13,530	13,530	13,530	13,530	13,530	13,530
Cuba	13,530	13,530	13,530	13,530	13,530	13,530
Cyprus	718	859	784	855	1,003	1,044
Czech & Slovak Republics	31,430	31,970	33,770	32,370	33,450	33,470
Denmark	31,870	35,000	36,970	39,430	40,890	41,790
Djibouti	no data					
Dominican Republic	961	902	863	755	827	791
Ecuador	837	1,186	1,398	1,362	1,130	1,161
Egypt	23,000	27,350	26,470	35,230	32,930	34,380
El Salvador	1,082	1,144	1,229	1,140	1,010	1,138
Equatorial Guinea	27	27	27	27	27	27
Ethiopia	1,096	1,245	1,306	1,457	2,018	1,655
Fiji	285	282	339	330	309	328
Finland	24,140	25,220	25,510	27,360	29,020	28,440
Former Soviet Union	509,600	546,600	571,300	616,400	606,800	615,500
France	302,600	314,200	340,200	368,600	370,200	376,200
Gabon	1,486	1,150	1,304	1,299	1,407	1,298
Germany (FRG)	289,000	307,900	318,800	320,500	319,800	331,400
Germany (GDR)	60,480	63,470	65,790	70,210	73,680	76,870
Ghana	666	474	453	430	304	407
Greece	15,410	16,530	18,690	24,480	20,640	24,210
Guatemala	898	1,194	1,323	1,141	982	811
Guinea	1,252	1,162	1,234	1,169	1,063	750

Country	Central Government Expenditures (US$ M 1989)				
	1985	1986	1987	1988	1989
Afghanistan	*508*	*508*	*508*	*508*	*508*
Albania	1,491	*1,491*	1,463	1,434	*1,434*
Algeria	18,840	18,850	17,790	17,480	*17,480*
Angola	*2,700*	*2,700*	*2,700*	3,322	*3,322*
Argentina	17,000	14,470	14,540	9,364	*9,364*
Australia	70,320	72,050	72,040	69,160	66,620
Austria	45,390	47,170	47,340	49,320	49,340
Bahamas					
Bahrain	1,598	1,511	1,534	967	1,491
Bangladesh	2,294	2,200	2,200	2,200	2,139
Belgium	77,760	75,870	74,540	75,860	53,690
Belize					
Benin	*350*	*350*	303	202	168
Bolivia	*1,100*	611	573	653	717
Botswana	743	812	1,013	1,138	1,035
Brazil	154,600	160,200	212,000	256,000	*256,000*
Brunei					
Bulgaria	20,320	23,850	23,540	19,690	19,800
Burkina Faso	212	267	281	307	*307*
Burundi	129	136	156	166	195
Cambodia	*269*	*269*	*269*	*269*	*269*
Cameroon	2,849	2,993	2,876	*2,500*	2,232
Canada	118,000	114,800	115,800	121,400	123,000
Cape Verde	*200*	*200*	*200*	*200*	278
Central African Republic	296	278	283	268	274
Chad	49	86	90	83	*83*
Chile	4,417	4,322	4,439	6,600	*6,600*
China	94,900	113,800	113,100	113,900	117,100
Colombia	4,798	4,642	5,012	4,454	4,627
Congo	909	1,136	817	861	752
Costa Rica	969	1,223	1,325	1,249	1,394
Cote d'Ivoire	*13,530*	*13,530*	*13,530*	*13,530*	13,530
Cuba	*13,530*	*13,530*	*13,530*	*13,530*	13,530
Cyprus	1,078	1,104	1,175	1,269	1,345
Czech & Slovak Republics	34,390	34,520	35,490	38,260	38,540
Denmark	41,570	39,780	39,700	41,380	42,090
Djibouti					
Dominican Republic	817	840	1,144	1,219	1,043
Ecuador	1,397	1,511	1,375	1,368	1,402
Egypt	34,900	36,730	32,100	33,760	32,590
El Salvador	1,082	845	783	709	675
Equatorial Guinea	*27*	*27*	*27*	27	27
Ethiopia	1,857	1,914	1,991	2,242	2,507
Fiji	312	335	317	286	*286*
Finland	30,220	31,630	32,910	33,710	33,850
Former Soviet Union	630,700	680,300	709,900	712,500	680,000
France	382,900	389,600	386,800	401,300	410,500
Gabon	1,389	2,167	*1,600*	1,108	1,127
Germany (FRG)	332,900	332,700	339,200	351,400	353,200
Germany (GDR)	83,270	83,270	86,660	91,540	93,780
Ghana	588	638	671	703	*703*
Greece	25,210	18,440	19,390	21,240	23,090
Guatemala	713	736	874	986	1,007
Guinea	*750*	450	532	599	549

Country	Central Government Expenditures (US$ M 1989)					
	1979	1980	1981	1982	1983	1984
Guinea-Bissau	68	79	68	71	63	79
Guyana	271	333	368	368	368	368
Haiti	407	405	455	541	565	436
Honduras	744	867	812	888	869	986
Hungary	29,070	27,630	30,010	29,020	30,300	29,680
India	28,610	30,890	31,670	34,540	37,080	41,980
Indonesia	11,240	12,690	14,950	13,630	14,730	13,800
Iran	21,700	19,680	19,900	21,700	18,480	17,420
Iraq	35,000	48,300	37,240	33,280	33,280	33,280
Ireland	11,920	13,490	14,640	15,790	15,690	15,780
Israel	25,230	26,630	28,960	28,540	36,110	35,290
Italy	241,332	260,610	292,332	308,652	338,742	353,124
Jamaica	1,235	1,312	1,369	1,363	1,247	1,198
Japan	372,300	397,392	415,140	425,544	440,130	446,046
Jordan	2,274	2,320	2,341	2,402	2,285	2,239
Kenya	1,686	1,719	1,911	2,014	1,757	1,745
Kuwait	8,930	10,770	11,790	13,020	12,790	12,160
Laos	277	285	285	285	285	302
Lebanon	2,237	2,509	1,968	2,446	2,790	2,202
Lesotho	195	190	213	169	174	202
Liberia	503	396	445	478	457	382
Libya	12,830	17,120	15,000	13,720	16,170	14,790
Luxembourg	2,372	2,447	2,546	2,596	2,744	2,587
Madagascar	583	594	682	485	372	435
Malawi	426	485	415	350	355	357
Malaysia	5,618	7,660	10,780	11,500	10,320	9,868
Mali	221	348	377	489	514	508
Malta	500	505	594	645	624	620
Mauritania	339	394	388	371	293	260
Mexico	27,040	31,330	39,540	56,760	45,660	43,180
Mongolia	no data					
Morocco	5,226	5,398	6,183	6,520	5,584	5,345
Mozambique	303	303	327	312	260	260
Myanmar/Burma	1,955	2,287	2,456	2,695	2,621	2,708
Namibia	no data					
Nepal	243	258	285	343	399	394
Netherlands	100,500	105,200	108,900	111,100	114,100	115,700
New Zealand	14,260	14,720	15,870	16,100	16,660	17,610
Nicaragua	270	416	564	704	1,010	934
Niger	363	439	422	490	354	314
Nigeria	5,081	4,860	4,471	4,227	3,909	3,444
North Korea	14,025	14,025	14,025	14,025	14,025	14,025
North Yemen	1,320	1,366	1,836	2,100	1,974	1,754
Norway	29,250	29,090	27,840	28,360	29,780	31,000
Oman	2,373	3,142	3,767	4,021	4,323	4,760
Pakistan	4,800	4,848	5,311	5,174	6,099	6,493
Panama	1,538	1,514	1,695	2,019	1,834	1,904
Papua New Guinea	983	1,056	1,188	1,130	1,035	1,014
Paraguay	327	357	424	401	375	390
Peru	7,079	10,080	10,160	9,550	8,959	8,676
Philippines	4,947	5,448	6,222	6,279	5,700	4,781
Poland	57,440	61,180	62,200	42,060	36,680	40,610
Portugal	12,930	14,600	16,270	16,100	16,790	15,950
Qatar	3,652	4,427	5,444	4,379	4,500	4,670
Romania	51,370	43,840	38,490	29,970	25,720	27,530

	Central Government Expenditures (US$ M 1989)				
Country	1985	1986	1987	1988	1989
Guinea-Bissau	89	56	83	76	76
Guyana	368	368	368	368	368
Haiti	450	392	431	431	431
Honduras	989	894	848	826	826
Hungary	28,060	27,490	26,780	27,320	20,260
India	48,370	53,720	54,340	59,140	60,120
Indonesia	15,910	18,740	16,950	17,380	18,340
Iran	13,790	13,510	12,980	12,980	12,980
Iraq	33,280	33,280	33,280	33,280	33,280
Ireland	16,330	15,920	16,240	13,020	12,230
Israel	29,420	25,520	25,880	25,230	22,840
Italy	360,570	401,064	417,384	436,764	436,764
Jamaica	1,247	1,090	1,200	1,439	1,439
Japan	454,920	460,326	475,422	490,314	490,314
Jordan	2,482	2,474	2,776	2,739	2,739
Kenya	1,889	1,884	2,285	2,096	2,096
Kuwait	11,920	11,240	9,687	9,560	9,875
Laos	302	302	302	302	302
Lebanon	1,700	1,700	1,273	1,273	1,273
Lesotho	186	165	165	147	193
Liberia	372	346	311	248	435
Libya	13,000	11,960	11,960	11,960	11,330
Luxembourg	2,630	2,725	2,844	3,004	3,004
Madagascar	517	502	300	168	122
Malawi	436	470	428	278	278
Malaysia	11,400	11,750	9,755	10,260	10,260
Mali	571	588	545	545	545
Malta	698	659	691	708	708
Mauritania	237	275	275	275	301
Mexico	47,440	53,200	57,890	52,010	38,420
Mongolia					
Morocco	5,844	5,989	5,615	5,615	5,615
Mozambique	204	220	230	236	236
Myanmar/Burma	2,800	2,845	2,326	1,847	1,847
Namibia					
Nepal	422	443	467	538	535
Netherlands	114,000	111,400	116,500	118,600	119,600
New Zealand	17,520	18,100	17,870	16,310	15,150
Nicaragua	834	664	625	594	254
Niger	321	292	375	463	463
Nigeria	3,552	5,087	7,400	8,221	5,460
North Korea	14,025	14,025	14,515	14,647	14,647
North Yemen	1,532	1,487	1,865	2,223	2,076
Norway	32,700	36,830	39,060	40,330	42,160
Oman	5,216	4,604	3,764	3,725	3,752
Pakistan	7,071	8,569	8,766	8,876	10,130
Panama	1,773	1,932	1,935	1,445	1,445
Papua New Guinea	1,014	1,082	1,054	1,038	1,192
Paraguay	357	312	362	368	368
Peru	8,448	9,118	9,545	6,395	5,554
Philippines	4,854	6,595	6,664	6,890	7,912
Poland	41,050	40,240	35,690	35,160	66,160
Portugal	18,360	19,200	18,660	19,080	19,440
Qatar	4,895	5,105	3,620	3,539	3,539
Romania	39,010	41,510	37,470	39,120	40,970

Country	Central Government Expenditures (US$ M 1989)					
	1979	1980	1981	1982	1983	1984
Rwanda	247	271	297	319	347	375
Saudi Arabia	72,900	82,620	87,940	100,500	101,800	82,450
Senegal	707	831	925	1,126	1,122	1,180
Seychelles	no data					
Sierra Leone	199	202	199	164	170	126
Singapore	3,251	3,768	4,925	5,001	5,921	5,351
Somali Republic	363	184	162	198	164	109
South Africa	19,790	19,580	21,850	21,900	22,800	24,340
South Korea	17,381	17,758	20,135	21,644	21,247	22,389
South Yemen	568	750	750	750	989	950
Spain	79,910	82,650	85,800	91,790	98,250	107,500
Sri Lanka	1,769	2,097	1,722	1,837	1,862	1,868
Sudan	2,786	2,684	2,883	2,976	2,800	2,800
Suriname	596	582	676	683	656	636
Sweden	67,150	68,440	72,130	73,860	75,950	75,900
Switzerland	29,470	30,110	29,660	30,330	31,350	32,660
Syria	5,789	7,966	6,980	8,391	9,906	9,742
Taiwan	15,190	16,950	15,000	13,180	14,090	13,330
Tanzania	799	669	638	728	6,115	547
Thailand	6,010	7,110	7,272	8,556	8,642	9,060
The Gambia	52	49	55	59	63	63
Togo	503	402	393	369	340	398
Trinidad and Tobago	2,125	2,227	2,692	3,346	3,903	2,469
Tunisia	2,603	2,599	2,730	3,162	3,458	3,612
Turkey	15,180	13,250	13,060	13,550	14,130	15,470
UAE	3,792	6,143	7,362	6,874	5,403	5,004
Uganda	214	186	197	435	444	333
United Kingdom	257,500	261,500	265,600	275,800	283,800	281,600
United States	843,500	916,900	966,000	992,300	1,041,000	1,051,000
Uruguay	1,682	1,873	2,167	2,358	1,879	1,739
Venezuela	8,560	9,607	12,990	11,910	9,467	8,863
Vietnam	6,547	6,547	6,547	6,547	6,547	6,547
Yugoslavia	5,299	5,064	4,703	4,289	4,366	4,244
Zaire	912	996	1,151	1,102	926	1,266
Zambia	1,395	1,985	1,751	1,995	1,519	1,403
Zimbabwe	1,234	1,491	1,504	1,963	1,807	2,109

Country	Central Government Expenditures (US$ M 1989)				
	1985	1986	1987	1988	1989
Rwanda	404	390	377	365	355
Saudi Arabia	89,960	59,960	37,650	39,220	38,100
Senegal	1,226	1,476	1,405	1,488	1,419
Seychelles					
Sierra Leone	104	85	85	69	69
Singapore	7,511	8,048	8,207	5,770	7,807
Somali Republic	112	116	116	116	116
South Africa	25,670	26,700	28,520	24,640	28,130
South Korea	23,582	24,245	26,602	29,121	29,121
South Yemen	950	950	950	907	907
Spain	108,800	110,000	118,500	103,800	135,500
Sri Lanka	2,145	2,170	2,161	2,345	2,194
Sudan	2,800	2,800	2,800	2,550	2,550
Suriname	674	713	614	550	543
Sweden	79,240	75,190	74,500	74,020	74,600
Switzerland	33,490	35,350	35,350	37,760	37,760
Syria	9,831	6,735	4,793	4,376	3,200
Taiwan	14,520	15,070	14,240	21,940	26,560
Tanzania	567	499	571	664	1,095
Thailand	10,470	10,300	10,160	9,973	10,410
The Gambia	63	63	63	63	74
Togo	425	461	388	280	265
Trinidad and Tobago	2,189	1,916	1,694	1,492	1,338
Tunisia	3,573	3,754	3,482	3,484	3,669
Turkey	16,210	14,860	16,550	16,650	18,320
UAE	4,943	4,041	3,884	3,739	3,613
Uganda	375	338	283	294	294
United Kingdom	292,500	292,300	289,100	282,400	285,700
United States	1,143,000	1,149,000	1,137,000	1,164,000	1,190,000
Uruguay	1,660	1,799	1,926	2,072	1,500
Venezuela	9,076	10,580	10,830	9,629	7,792
Vietnam	6,547	6,547	4,507	4,507	4,507
Yugoslavia	3,910	4,000	4,465	5,406	3,981
Zaire	1,045	1,062	1,400	1,715	1,517
Zambia	1,721	2,148	1,668	1,381	770
Zimbabwe	2,101	2,110	2,389	2,467	2,580

7. Global Military Expenditires, 1979–1989

Figures are in millions of FY89 dollars. Where estimates are employed, figures are denoted in italics. Estimates were typically interpolated between two known values.

Country	Military Expenditures (US$ M FY89)					
	1979	1980	1981	1982	1983	1984
Afghanistan	252	245	212	186	246	327
Albania	*180*	*180*	180	169	158	168
Algeria	1,361	1,385	1,859	1,802	1,383	1,456
Angola	*651*	*651*	*651*	*651*	651	*651*
Argentina	2,039	2,296	2,214	3,527	2,569	2,751
Australia	4,774	4,753	5,285	5,586	5,988	6,286
Austria	1,222	1,208	1,197	1,336	1,452	1,472
Bahamas	no data					
Bahrain	231	232	289	354	202	173
Bangladesh	208	215	238	269	351	323
Belgium	4,148	4,348	4,504	4,420	4,329	4,181
Belize	no data					
Benin	23	27	45	45	44	41
Bolivia	105	122	145	75	49	86
Botswana	36	35	34	30	34	45
Brazil	2,661	2,799	2,627	3,367	3,078	3,002
Brunei	no data					
Bulgaria	5,739	5,883	5,936	6,507	6,498	6,460
Burkina Faso	35	35	38	41	40	40
Burundi	20	28	32	28	30	27
Cambodia	no data					
Cameroon	107	116	97	104	212	214
Canada	7,036	7,626	7,477	8,377	8,413	9,415
Cape Verde	*18*	18	20	*20*	*20*	20
Central African Republic	20	20	21	20	18	*18*
Chad	25	20	*15*	*10*	8	10
Chile	622	662	723	677	667	693
China	26,950	24,120	23,640	23,780	23,220	22,600
Colombia	265	306	294	434	392	427
Congo	*51*	51	54	79	74	*79*
Costa Rica	29	28	25	23	30	31
Cote d'Ivoire	202	155	156	164	111	120
Cuba	1,865	1,680	1,613	1,680	1,788	1,625
Cyprus	51	38	55	51	52	50
Czech / Slovak Republics	7,393	7,734	7,858	8,562	8,763	8,835
Denmark	2,046	2,143	2,201	2,256	2,257	2,217
Djibouti	*50*	*50*	*50*	*50*	*50*	*50*
Dominican Republic	92	70	76	74	72	66
Ecuador	282	258	278	282	263	210
Egypt	5,187	4,306	4,240	7,820	7,384	8,024
El Salvador	157	177	213	232	230	353
Ethiopia	408	473	445	451	441	456
Fiji	10	9	11	14	14	15
Finland	1,127	1,139	1,351	1,554	1,601	1,605
Former Soviet Union	284,400	292,000	295,200	300,500	304,900	309,200
France	30,080	31,270	32,610	33,310	33,840	33,710
Gabon	82	97	78	82	75	63
Germany (FRG)	32,370	33,000	34,130	33,980	34,280	34,070
Germany (GDR)	10,990	11,170	11,520	12,360	12,680	13,010
Ghana	29	17	28	26	13	24
Greece	2,859	2,621	3,230	3,191	2,948	3,453

Country	Military Expenditures (US$ M FY89)				
	1985	1986	1987	1988	1989
Afghanistan	327	327	327	327	327
Albania	163	163	162	163	157
Algeria	1,183	1,411	1,359	1,660	2,313
Angola	800	1,250	1,250	1,250	1,250
Argentina	2,103	2,302	2,169	2,070	1,858
Australia	6,666	6,599	7,038	6,342	6,153
Austria	1,505	1,556	1,423	1,364	1,402
Bahamas					
Bahrain	171	178	173	195	196
Bangladesh	298	314	351	335	323
Belgium	4,089	4,163	4,172	3,968	3,881
Belize					
Benin	37	33	35	36	33
Bolivia	110	138	152	163	182
Botswana	43	60	110	119	62
Brazil	3,180	3,992	4,502	5,966	5,966
Brunei					
Bulgaria	6,612	6,897	6,938	6,109	5,885
Burkina Faso	40	59	49	54	54
Burundi	27	32	33	30	28
Cambodia					
Cameroon	238	270	241	166	148
Canada	10,180	10,550	10,800	10,930	10,840
Cape Verde	20	20	20	20	20
Central African Republic	18	18	18	18	18
Chad	16	32	36	36	36
Chile	704	662	882	836	790
China	22,600	21,970	22,030	22,720	22,330
Colombia	482	472	533	640	758
Congo	84	128	102	102	102
Costa Rica	27	27	27	22	22
Cote d'Ivoire	115	134	182	199	199
Cuba	1,520	1,450	1,405	1,405	1,377
Cyprus	44	31	37	43	41
Czech / Slovak Republics	9,105	9,388	9,508	9,505	8,361
Denmark	2,173	2,076	2,171	2,222	2,184
Djibouti	50	50	50	50	50
Dominican Republic	65	75	83	63	52
Ecuador	237	253	246	229	163
Egypt	7,724	7,168	7,167	5,786	3,499
El Salvador	314	298	277	247	252
Ethiopia	435	430	464	600	763
Fiji	13	13	24	25	26
Finland	1,633	1,784	1,693	1,756	1,788
Former Soviet Union	315,600	319,200	325,900	330,900	311,000
France	33,690	33,840	34,980	34,860	35,260
Gabon	92	166	137	170	140
Germany (FRG)	34,130	33,950	33,960	33,630	33,600
Germany (GDR)	13,280	13,690	14,150	14,320	14,320
Ghana	42	40	42	22	30
Greece	3,474	3,097	3,148	3,300	3,097

Country	Military Expenditures (US$ M FY89)					
	1979	1980	1981	1982	1983	1984
Guatemala	90	109	105	144	152	142
Guinea	126	126	126	100	84	82
Guinea-Bissau	5	6	5	6	10	5
Guyana	17	19	21	23	20	21
Haiti	27	38	43	39	34	35
Honduras	85	78	63	73	94	145
Hungary	4,017	4,173	4,129	4,407	4,412	4,326
India	5,132	5,216	5,795	6,431	6,911	7,353
Indonesia	1,538	1,611	1,813	1,830	1,632	1,713
Iran	5,446	3,867	4,119	4,405	4,000	4,000
Iraq	8,704	8,704	8,704	8,704	8,704	8,704
Ireland	376	421	414	501	469	464
Israel	9,218	9,805	8,281	7,611	8,420	9,225
Italy	13,811	14,515	14,729	15,800	16,136	16,657
Jamaica	26	26	34	44	37	22
Japan	19,207	19,706	20,553	21,746	22,981	24,184
Jordan	914	848	817	836	818	849
Kenya	272	263	191	249	237	219
Kuwait	1,231	1,314	1,153	1,414	1,701	1,676
Laos	80	80	80	80	80	64
Lebanon	532	560	464	404	558	507
Lesotho	10	12	14	14	13	15
Liberia	13	20	46	61	34	27
Libya	5,373	4,555	4,536	4,517	5,174	5,917
Luxembourg	50	58	60	60	62	63
Madagascar	83	80	71	55	38	45
Malawi	58	55	35	26	22	21
Malaysia	807	1,002	1,472	1,530	1,363	1,093
Mali	38	37	37	41	40	39
Malta	8	8	11	16	18	16
Mauritania	99	95	74	69	61	60
Mexico	723	671	901	866	868	1,137
Mongolia	257	296	266	250	244	229
Morocco	836	1,124	1,116	1,376	1,306	1,250
Mozambique	75	84	97	91	83	83
Myanmar/Burma	476	502	547	520	522	518
Namibia	no data					
Nepal	16	17	19	19	23	25
Netherlands	6,064	5,965	6,094	6,062	6,061	6,240
New Zealand	606	707	793	811	778	790
Nicaragua	39	78	100	151	159	195
Niger	16	16	16	14	15	13
Nigeria	663	658	612	501	435	390
North Korea	6,452	6,197	5,860	5,700	5,698	5,696
North Yemen	515	409	528	678	670	508
Norway	1,996	1,945	1,960	2,059	2,196	2,112
Oman	1,125	1,560	1,825	1,909	2,121	2,220
Pakistan	1,036	1,145	1,261	1,406	1,768	1,752
Panama	43		52	71	99	108
Papua New Guinea	40	46	46	45	41	47
Paraguay	44	48	56	64	84	56
Peru	1,503	2,147	1,926	2,113	1,899	2,301
Philippines	833	738	718	665	656	454
Poland	13,730	14,210	14,100	15,710	15,480	16,030
Portugal	1,189	1,265	1,276	1,285	1,250	1,195

Country	Military Expenditures (US$ M FY89)				
	1985	1986	1987	1988	1989
Guatemala	121	91	136	133	131
Guinea	55	55	55	30	30
Guinea-Bissau	4	4	3	3	3
Guyana	33	40	40	46	46
Haiti	34	40	42	34	34
Honduras	138	141	142	120	120
Hungary	4,305	4,345	4,361	4,576	4,064
India	7,584	8,103	8,900	8,721	8,174
Indonesia	1,646	1,739	1,477	1,460	1,510
Iran	4,000	4,000	4,000	4,000	4,000
Iraq	8,704	8,704	8,704	8,704	8,704
Ireland	477	473	446	448	449
Israel	7,999	7,137	6,269	5,814	5,745
Italy	17,330	17,411	19,533	20,839	20,839
Jamaica	23	25	28	32	32
Japan	25,500	26,663	28,101	29,447	29,447
Jordan	877	647	622	900	900
Kenya	161	167	206	294	294
Kuwait	1,736	1,443	1,358	1,325	1,964
Laos	63	63	63	63	63
Lebanon	507	507	507	507	507
Lesotho	17	19	16	15	14
Liberia	33	41	44	43	58
Libya	5,000	4,000	2,984	3,100	3,309
Luxembourg	62	64	72	82	76
Madagascar	41	41	40	33	35
Malawi	25	31	24	34	34
Malaysia	1,062	1,223	1,350	908	908
Mali	46	48	46	45	45
Malta	18	19	22	22	22
Mauritania	59	53	37	38	40
Mexico	1,194	1,073	1,008	1,002	875
Mongolia	219	276	295	293	259
Morocco	1,250	1,196	1,185	1,147	1,203
Mozambique	77	78	79	96	107
Myanmar/Burma	550	528	518	454	611
Namibia					
Nepal	26	29	30	31	33
Netherlands	6,185	6,298	6,398	6,319	6,399
New Zealand	790	854	827	872	847
Nicaragua	218	218	218	218	218
Niger	16	17	18	19	27
Nigeria	335	270	198	210	130
North Korea	5,750	5,796	5,826	5,840	5,840
North Yemen	432	408	408	668	618
Norway	1,450	2,583	2,820	2,790	2,925
Oman	2,205	1,920	1,632	1,405	1,552
Pakistan	1,990	2,149	2,257	2,404	2,488
Panama	107	109	107	105	141
Papua New Guinea	46	47	47	48	48
Paraguay	43	42	40	57	61
Peru	2,682	3,109	2,635	2,635	2,635
Philippines	461	695	697	702	960
Poland	16,700	16,900	16,620	15,980	15,480
Portugal	1,196	1,248	1,272	1,393	1,457

Country	Military Expenditures (US$ M FY89)					
	1979	1980	1981	1982	1983	1984
Qatar	763	891	891	891	891	891
Romania	6,835	6,759	6,704	7,178	7,628	7,637
Rwanda	30	35	34	42	49	44
Saudi Arabia	19,910	22,090	24,750	27,830	30,160	23,910
Senegal	127	139	108	110	108	103
Seychelles	no data					
Sierra Leone	10	7	9	8	7	5
Singapore	687	785	868	895	819	1,119
Somali Republic	62	34	40	40	37	30
South Africa	3,038	3,777	3,337	2,772	3,564	3,304
South Korea	4,543	5,103	5,484	5,766	5,814	5,833
South Yemen	140	149	181	198	208	230
Spain	6,145	6,719	6,872	6,998	7,295	7,231
Sri Lanka	72	72	65	67	83	91
Sudan	449	445	426	558	280	277
Suriname	34	34	34	34	34	34
Sweden	4,880	4,799	4,870	5,037	5,057	4,977
Switzerland	3,612	3,626	3,626	3,723	3,794	3,890
Syria	2,352	2,852	2,634	2,891	4,061	4,059
Taiwan	4,539	4,748	4,962	6,171	6,815	6,244
Tanzania	207	61	76	91	83	74
Thailand	1,421	1,489	1,482	1,651	1,655	1,796
The Gambia	1	1	1	1	1	1
Togo	27	27	28	26	23	25
Trinidad and Tobago	47	32	41	123	177	146
Tunisia	352	288	214	317	389	270
Turkey	2,234	2,409	2,615	2,924	2,818	2,722
UAE	1,924	2,521	2,809	2,500	2,399	2,265
Uganda	41	45	60	71	64	55
United Kingdom	30,420	32,530	30,730	32,560	35,600	36,100
United States	210,012	214,198	239,346	267,054	287,790	302,499
Uruguay	188	246	334	312	233	189
Venezuela	543	467	417	607	750	680
Vietnam	2,670	2,670	2,670	2,670	2,670	2,670
Yugoslavia	2,400	2,459	2,372	2,152	2,030	2,112
Zaire	100	85	45	87	125	173
Zambia	520	607	450	328	300	276
Zimbabwe	254	371	289	307	302	316

Country	Military Expenditures (US$ M FY89)				
	1985	1986	1987	1988	1989
Qatar	891	891	891	891	891
Romania	7,787	7,833	8,050	7,291	6,916
Rwanda	38	41	45	37	33
Saudi Arabia	24,290	19,190	17,430	14,160	14,690
Senegal	108	110	94	92	90
Seychelles					
Sierra Leone	5	5	6	5	5
Singapore	1,281	1,213	1,223	1,398	1,475
Somali Republic	33	35	35	35	35
South Africa	2,974	3,074	3,574	3,701	3,786
South Korea	6,154	6,526	6,639	7,202	7,202
South Yemen	225	234	223	226	226
Spain	7,558	7,219	8,125	7,541	7,775
Sri Lanka	179	159	203	313	223
Sudan	250	223	337	298	339
Suriname	36	35	37	40	39
Sweden	4,998	4,935	4,991	5,075	4,872
Switzerland	3,970	3,898	3,610	3,610	3,806
Syria	4,129	3,230	1,937	1,875	2,234
Taiwan	7,260	7,535	5,913	6,877	8,060
Tanzania	73	78	84	97	110
Thailand	2,065	1,920	1,860	1,813	1,843
The Gambia	1	1	1	1	1
Togo	29	35	43	43	43
Trinidad and Tobago	100	100	100	100	59
Tunisia	315	313	289	247	273
Turkey	2,906	3,340	3,205	2,942	3,150
UAE	2,165	1,753	1,710	1,652	1,471
Uganda	58	89	74	61	60
United Kingdom	37,070	36,380	35,840	34,050	34,630
United States	313,678	308,066	303,193	298,746	290,919
Uruguay	176	181	148	169	169
Venezuela	407	544	1,298	674	407
Vietnam	2,670	2,670	2,670	2,670	2,670
Yugoslavia	2,144	2,398	2,460	2,631	2,126
Zaire	103	193	225	242	242
Zambia	200	150	125	109	65
Zimbabwe	303	330	386	369	386

Selected Bibliography

Ackerman, Julia A., and Michael Dunn, "Chinese Airpower Revs Up," *Air Force Magazine*, July 1993.

"The Air Force Today and Tomorrow," interview with Colonel General Yevgeniy I. Shaposhnikov, *Aviatsiia i kosmonavtika*, No. 8, August 1990.

Arms Control and Disarmament Agency, *World Military Expenditures and Arms Transfers 1990 (WMEAT)*, Washington, D.C.: U.S. Government Printing Office, 1990.

Aspin, Les, *Report on the Bottom-Up Review*, Washington, D.C.: Department of Defense, October 1993.

Aviation Advisory Services Limited, *International Air Forces & Military Aircraft Directory*, 1991.

Aviation Week and Space Technology, selected issues.

Barrie, Douglas, and Alexander Velovich, "Fighting Over the Scraps," *Flight International*, June 1–7, 1994.

Betts, Richard K. (ed.), *Cruise Missiles: Technology, Strategy, Politics*, Washington, D.C.: The Brookings Institution, 1981.

Bitzinger, Richard, *Facing the Future: The Swedish Air Force, 1990–2005*, Santa Monica, CA: RAND, R-4007-RC, 1991.

Bowie, Christopher, Fred Frostic, Kevin Lewis, John Lund, David Ochmanek, and Philip Propper, *The New Calculus: Analyzing Airpower's Changing Role in Joint Theater Campaigns*, Santa Monica, CA: RAND, MR-149-AF, 1993.

Bowie, Christopher, Mark Lorell, and John Lund, *Trends in NATO Central Region Tactical Fighter Inventories 1950–2005*, Santa Monica, CA: RAND, N-3053-AF, 1990.

Chant, Christopher, *World Encyclopedia of Modern Air Weapons*, Wellingsborough: Patrick Stevens Limited Publishing, 1988.

"China's Military Aircraft Inventory," *Flight International*, November 25, 1992.

Cohen, Eliot A., "The Mystique of U.S. Air Power," *Foreign Affairs*, Vol. 73, No. 1, January/February 1994.

Cohen, Elizier, *Israel's Best Defense: The First Full Story of the Israeli Air Force*, New York: Orion Books, 1993.

Cohen, Jon, "China's Combat Aircraft Domestic and Export Production Schedules for Firm Orders 1992–2000," *Global Arms Market*, Table 4.12, Institute for Defense and Disarmament Studies, 1994.

Davis, Paul K. (ed.), *New Challenges for Defense Planning: Rethinking How Much Is Enough*, Santa Monica, CA: RAND, MR-400-RC, 1994.

Defense and Foreign Affairs Handbook, Washington, D.C.: Perth, 1990–1991.

Defense and Foreign Affairs Strategic Policy, Washington, D.C.: Copley & Associates, December 31, 1992.

Department of Defense, *Conduct of the Persian Gulf War*, Washington, D.C., April 1992.

"Flying in the Face of Adversity," interview with Petr Butowski, *Jane's Defense Weekly*, April 17, 1993.

Frostic, Fred, *Air Campaign Against the Iraqi Army in the Kuwaiti Theater of Operations*, Santa Monica, CA: RAND, MR-357-AF, 1994.

Frostic, Fred, and Christopher J. Bowie, "Conventional Campaign Analysis of Major Regional Conflicts," in Davis (1994).

Garrity, Patrick J., *Why the Gulf War Still Matters: Foreign Perspectives on the War and the Future of International Security*, Los Alamos: Center for National Security Studies, Los Alamos National Laboratory, Report No. 16, July 1993.

Harshberger, Edward R., *Long Range Conventional Missiles: Issues for Near-Term Development*, Santa Monica, CA: RAND, N-3328-RGSD, 1991.

Harvey, John R., "Regional Ballistic Missiles and Advanced Strike Aircraft: Comparing Military Effectiveness," *International Security*, Vol. 17, No. 2, Fall 1992.

Hosmer, Stephen T., "Weapons of Mass Destruction and the Persian Gulf War," *Project AIR FORCE Annual Report Fiscal Year 1993*, Santa Monica, CA: RAND, AR-3900-AF, 1994, pp. 9–12.

International Defense Review, selected issues.

International Institute for Strategic Studies, *The Military Balance*, London: IISS, selected issues.

International Monetary Fund, *International Financial Statistics*, Washington, D.C.: IMF, 1992.

Irving, David, *The Mare's Nest*, London: William Kimber and Co., 1964.

Jane's Defence Weekly, selected issues.

Jane's Information Group, *DMS Market Intelligence Reports, Foreign Military Markets NATO & Europe*, London, 1989.

———, *Jane's All the World's Aircraft*, London, 1991–1992.

———, *Jane's Land-Based Air Defence 1992–93*, London, 1992–93.

Jones, Greg, *The Iraqi Ballistic Missile Program: The Gulf War and the Future of the Missile Threat*, Marina del Rey, CA: American Institute for Strategic Cooperation, 1992.

Kaminer, R. A., "Israel Reveals Unprecedented Level of Defense-Budget Details," *International Defense Review*, January 1994.

Keaney, Thomas A., and Eliot A. Cohen, *Gulf War Air Power Survey Summary Report*, Washington, D.C.: U.S. Government Printing Office, 1993.

Lambeth, Benjamin, *Desert Storm and Its Meaning: The View From Moscow,* Santa Monica, CA: RAND, R-4164-AF, 1992.

———, "Russian Airpower at the Crossroads," *Project AIR FORCE Annual Report Fiscal Year 1993*, Santa Monica, CA: RAND, AR-3900-AF, 1994, pp. 50–53.

Large, J. P., A. A. Barbour, and G. F. Mills, *Procedures for Estimating Life-Cycle Costs of Electronic Combat Equipment,* Santa Monica, CA: RAND, R-3530-AF, 1988.

Levin, Mikhail, "That Same MiG," *Krylia Rodiny*, No. 3, 1992.

Lewis, Kevin, *Planning Future U.S. Fighter Forces*, Santa Monica, CA: RAND, MR-285-AF, 1993.

———, *The U.S. Air Force Budget and Posture over Time*, Santa Monica, CA: RAND, R-3807-AF, 1990.

Lorell, Mark, *The Future of Allied Tactical Fighter Forces in NATO's Central Region*, Santa Monica, CA: RAND, R-4144-AF, 1992.

McCormick, Gordon, *Stranger Than Fiction: Soviet Submarine Operations in Swedish Waters*, Santa Monica, CA: RAND, R-3776-AF, 1990.

Military Aviation News, selected issues.

Ministry of Defence, *RAF Air Power Doctrine (AP 3000)*, London: Ministry of Defence, 1992.

Millot, Dean, Roger Molander, and Peter Wilson, *"The Day After..." Study: Nuclear Proliferation in the Post–Cold War World*, 3 vols., Santa Monica, CA: RAND, MR-266-AF, 1993.

Missile Forecast, selected issues.

Missile Monitor, selected issues.

Munitions Systems Division, *1989 Weapons File*, Eglin AFB: United States Air Force, 1989.

Murray, Williamson, *Luftwaffe*, Baltimore: Nautical and Aviation Publishing Co., 1985.

Nagler, Robert, *Ballistic Missile Proliferation: An Emerging Threat*, Washington, D.C.: System Planning Corporation, 1992.

Nation, Joseph E., *German, British, and French Military Requirements and Resources to the Year 2005*, Santa Monica, CA: RAND, N-2982-RGSD, 1992.

Naval Institute Press, *Combat Fleets of the World, 1988/89, Their Ships, Aircraft, and Armament*.

Northrop Corporation (Advanced Technology and Design Center), *Air-To-Surface Munitions Handbook*, revision, 1992.

Office of Air Force History, *Encyclopedia of U.S. Air Force Aircraft and Missile Systems*, Washington, D.C., 1978.

Oxford University Press, *SIPRI Yearbook, 1992—World Armaments and Disarmament*, Oxford, 1992.

"Russia to Scrap 2,000 Aircraft," *Flight International*, March 31–April 6, 1993.

Shaver, Russell, Edward R. Harshberger, and Natalie W. Crawford, *Modernizing Airpower Projection Capabilities: Future Needs and Options*, Santa Monica, CA.: RAND, IP-126, 1993.

Simon, Yolande, *Prospects for the French Fighter Industry in a Post–Cold War Environment: Is the Future More Than a Mirage?* Santa Monica, CA: RAND, RGSD-106, 1993.

Spick, Michael, *Jet Fighter Performance: Korea to Vietnam*, London: Ian Alled Ltd., 1986.

Spellman, A., "Avionics Programs at Core of Latest Israeli Outreach to India," *Armed Forces Journal International*, November 1992.

Stanley, William, and Gary Liberson, *Measuring Effects of Payload and Radius Differences of Fighter Aircraft*, Santa Monica, CA: RAND, DB-102-AF, 1993.

Stein, David, *The Development of NATO Tactical Air Doctrine, 1970–1985*, Santa Monica, CA: RAND, R-3385-AF, 1987.

Tirpak, John, "The Secret Squirrels," *Air Force Magazine*, Vol. 77, No. 4, April 1994.

Terekhov, Vyacheslav, and Viktor Akimov, "Interview with Kokoshin," *Interfax*, December 3, 1992.

United States Air Force (SAF/OSX), *B-2 Survivability Against Air Defense Systems*, Washington, D.C.: Headquarters, United States Air Force, March 1990a.

United States Air Force, *Basic Aerospace Doctrine of the United States Air Force*, vol. 1, Headquarters, United States Air Force, 1990b.

"U.S., French Fighter Sales to Taiwan Nudge Mainland China Closer to Russia," *Armed Forces Journal International*, January 16, 1993.

Vladykin, O., "The Military Budget: Priority for Social Needs," interview with Lieutenant General V. Vorobyev, *Krasnaia zvezda*, February 4, 1992.

Werrell, Kenneth P., *Archie, Flak, AAA, and SAM, A Short Operational History of Ground-Based Air Defense*, Maxwell AFB, Alabama: Air University Press, 1988.

———, *The Evolution of the Cruise Missile*, Maxwell AFB, Alabama: Air University Press, 1985.

Wollen, M.S.D., "MiG-21 BIS Upgrade," *Indian Aviation*, October 1992.

Woodward, Sandy, *One Hundred Days: The Memoirs of the Falklands Battle Group Commander*, Annapolis: Naval Institute Press, 1992.

World Defense Almanac, 1991–1992, Baltimore, MD: Military Publication Service, 1991.

World Weapon Database, Vol 2, Soviet Military Aircraft, Lexington, MA: Heath, 1986.

Wright, Barton, *World Weapon Database, Vol. 1, Soviet Missiles*, Lexington, MA: Heath, 1986.

Yonay, Ehud, *No Margin for Error: The Making of the Israeli Air Force*, New York: Pantheon Books, 1993.